P9-BXX-919

THINGS THAT MAKE US SMART

THINGS THAT MAKE US SMART

DEFENDING HUMAN ATTRIBUTES IN THE AGE OF THE MACHINE

Donald A. Norman

A William Patrick Book

Addison-Wesley Publishing Company

Reading, Massachusetts Menlo Park, California New York
Don Mills, Ontario Wokingham, England Amsterdam Bonn
Sydney Singapore Tokyo Madrid San Juan
Paris Seoul Milan Mexico City Taipei

Library of Congress Cataloging-in-Publication Data

Norman, Donald A.
Things that make us smart: defending human attributes in the age of the machine / Donald A. Norman.
p. cm.
"A William Patrick book."
Includes bibliographical references and index.
ISBN 0-201-58129-9
1. Technology—Philosophy. 2. Cognitive science. 3. Man-machine systems. I. Title.
T14.N67 1993
303.48'34—dc20 92-36016
CIP

Jacket design by Caryl Hull Design Group
Text design by Jean Hammond
Set in 10½-point Trump Medieval by Shepard Poorman Communications

1 2 3 4 5 6 7 8 9-MA-9796959493
First printing, March 1993

To Julie, who makes me smart

Science Finds,
Industry Applies,
Man Conforms

Motto of the 1933 Chicago World's Fair

People Propose,
Science Studies,
Technology Conforms

A person-centered motto for the twenty-first century

CONTENTS

PREFACE

SOCIETY HAS UNWITTINGLY FALLEN INTO A MACHINE-CENTERED ORIENTA-tion to life, one that emphasizes the needs of technology over those of people, thereby forcing people into a supporting role, one for which we are most unsuited. Worse, the machine-centered viewpoint compares people to machines and finds us wanting, incapable of precise, repetitive, accurate actions. Although this is a natural comparison, and one that pervades society, it is also a most inappropriate view of people. It emphasizes tasks and activities that we should not be performing and ignores our primary skills and attributes—activities that are done poorly, if at all, by machines. When we take the machine-centered point of view, we judge things on artificial, mechanical merits. The result is continuing estrangement between humans and machines, continuing and growing frustration with technology and with the pace and stress of a technologically centered life.

It doesn't have to be this way. Today we serve technology. We need to reverse the machine-centered point of view and turn it into a person-centered point of view: Technology should serve us. This is as much a social problem as a technological one. The more I have interacted with the high-tech industries of the world, the more I have come to see that it is primarily our social structures that determine both the direction that technology takes and its impact upon our lives. That's why I call it a social problem, and that's what this book is about.

A QUARTET OF BOOKS

This book is the fourth in a series. *User Centered System Design* (UCSD) reported the outcome of studies performed by my research group, all of us trying to understand why computer and other systems were so difficult to use, so poorly designed. In *The Psychology of Everyday Things* (POET), I demonstrated that the same principles presented in UCSD for complex systems such as commercial aircraft, nuclear power plants, and computer systems also applied to mundane things such as doors, water faucets, and light switches. POET emphasized the everyday implications of design for the individual. The third book in the quartet, *Turn Signals Are the Facial Expressions of Automobiles,* stressed the social impact of our technologies. *Turn Signals* was intended to be a lighthearted and amusing treatment of important topics. This book, *Things That Make Us Smart,* is its complement, treating many of the same issues, but more deeply, more seriously. *Turn Signals* and *Things That Make Us Smart* make a good pair, each informing the other.

A PERSONAL NOTE

I have been increasingly bothered by the lack of reality in academic research. University-based research can be clever, profound, and deep, but surprisingly often it has little or no impact either upon scientific knowledge or upon society at large. University-based science is meant to impress one's colleagues: What matters is precision, rigor, and reproducibility, even if the result bears little relevance to the phenomena under study. Whether the work has any relevance to broader issues is seldom addressed. This is a common problem in the human and social sciences, where the phenomena are especially complex, variable, and heavily influenced by context. Most academic study is designed to answer questions raised by previous academic studies. This natural propensity tends to make the studies ever more specialized, ever more esoteric, thereby removed even further from concerns and issues of the world.

I worry about how much this criticism applies to my own work. I have long argued the importance of a person-centered design philosophy, but my work has also been academic, removed from real application. It is time to put my words into deeds. Therefore, during the writing of this book, I decided to leave my university and join the computer industry, the better to help guide products toward a more humane technology. This is especially important today, as our information technologies affect all aspects of life, where the products will look as unlike computers as the kitchen blender, clock, and dishwasher look unlike electric motors. This puts me to the test: In the real world of business, where many different practical constraints affect the final product, can a person-centered design philosophy have an effect? I will strive to see that it does.

ACKNOWLEDGMENTS

Many people have contributed to *Things That Make Us Smart.* Early versions of the book were distributed to many of my friends and used in my classes. The comments from my graduate students and colleagues have helped in the continuing succession of revisions. This final version of the book doesn't look very much like the early versions, in part due to the very useful feedback.

In particular, I thank Tom Erickson, Bill Gaver, Peter Hart, James Laffey, Wendy MacKay, James Spohrer, and Hank Strub. Jo-Anne Yates of MIT has taught me about the history of cognitive technology, especially as applied to business. Gerhard Fischer's "Topics in Symbolic AI" course at the University of Colorado, Boulder, provided a particularly provocative set of questions and discussions during my visit there.

Sandy Dijkstra has played an invaluable supportive role as my literary agent. Available when I needed her, encouraging when I needed encouragement. And of course, there is Julie, who has shared it all.

Bill Patrick, my editor at Addison-Wesley, continually prodded and poked me through the multiple versions of the book. Among other things, he fiercely defended the art of reading against my claim that the same or even enhanced benefits could come through properly constructed video (an interactive, computer-controlled video, to be sure). We carried out a prolonged debate through marginal notes and letters, causing me to examine more deeply the bases for my arguments, thereby strengthening and enriching them.

CHAPTER NOTES AND BOOK DESIGN

NOTHING SEEMS TO CREATE MORE CONTROVERSY ABOUT THE DESIGN OF A book than the placement of notes for each chapter. Academic readers are used to seeing notes at the bottom of relevant pages as footnotes. Trade publishers do not approve. They feel that notes distract, breaking the flow of reading. They prefer to hide the notes at the end of the book—out of sight, but still available for the serious reader.

Many of my readers have complained vociferously. The notes are hard to find, they say, and it is particularly disruptive to keep two place markers, one for text, one for notes. Alas, I know of no data relevant to the relative proportion of readers who might fall into each category, the one described by the publishers or the one corresponding to those who write me. Each side argues that the other constitutes a tiny minority.

For this book, we are trying a compromise that seemed to work for my previous book, *Turn Signals Are the Facial Expressions of Automobiles.* Most substantive comments have been put directly in the text; notes are used primarily for references and acknowledgments. I have kept substantive comments to a minimum in the notes, which should minimize the need to turn to them. Each note specifies the page from which it has come, with a short quote from the text indicating the section to which it refers. This makes it easy to go from the notes back to the text, a feature many readers requested.

There still remains the question of how to tell the reader that there is a relevant note at the back of the book. Traditional footnote symbols are thought to be disruptive, diverting the reader's attention to the note, usually to find that it does not contain essential information. We are trying a more subtle signal. Simple references and citations to statements made in the text are not marked, but if you wish to know the source for any such statement, look in the notes section at the end of the book. Substantive notes and discussions are signaled by the symbol * in the text.

ONE

A HUMAN-CENTERED TECHNOLOGY

THE GOOD NEWS IS THAT TECHNOLOGY CAN MAKE US SMART. IN FACT, IT already has. The human mind is limited in capability. There is only so much we can remember, only so much we can learn. But among our abilities is that of devising artificial devices—artifacts—that expand our capabilities. We invent things that make us smart. Through technology, we can think better and more clearly. We have access to accurate information. We can work effectively with others, whether together in the same place or separated in space or time. Three cheers for the invention of writing, reading, art, and music. Three cheers for the development of logic, the invention of encyclopedias and textbooks. Three cheers for science and engineering. Maybe.

The bad news is that technology can make us stupid. The technology for creating things has far outstripped our understanding of them. Things that make us smart can also make us dumb. They can entrap us with their seductive powers—as with commercial television—or frustrate us through their artificial complexity. Television has the power to entertain, we are told, but does it? Peer into the nation's living room in the evening and see the stultified masses, glued to their television sets. Television has the power to inform, we are told. The average television viewer in the United States watches twenty-one thousand commercials per year. News of the world is reduced to a few minutes per topic, each starting with the obligatory "reporter on the scene," where the reporter stands in front of some "local color," a background scene with appropriate sound and view to establish the fact that the reporter is actually there, followed by a few hundred words of text. A few hundred words of text. That is less than is on this single page. It would be only a tiny newspaper article, and even long newspaper articles are but brief summaries of complex events. Is this the kind of informing we had in mind?

Companies and homes are inundated with new artifacts, new technologies. The artifacts come with buttons, lights, and display screens. The old days of audiovisual presentations have given way to the new days of interactive multimedia in which computers merge with television, merge with the telephone. New forms of media are being invented, new methods of communication, of education, of entertainment, and of doing business. New consumer goods invade the home. All the ailments of the past, it is claimed, will disappear once these new approaches are in place. Are you dissatisfied with the school system? Don't worry, for now the science museums and school systems can use interactive, multimedia video computers and other fancy technologies to entrance the student. Every home will want to have one. Every office. Entrance the student? Entertain? Does entertainment make us smart?

I am a cognitive scientist, interested in the workings of the mind. My most recent research has concentrated upon the development of tools that aid the mind—mental tools I call "cognitive artifacts." My original goal in writing this book was to discuss how these tools work, what their principles were in adding to our mental abilities. Along the way, however, my studies caused me to question the manner by which our cognitive abilities are, in turn, manipulated by the tools cognition has helped to create.

We humans have invented a wide variety of things to aid our cognition, some physical, some mental. Thus tools such as paper, pencils, calculators, and computers are physical artifacts that aid cognition. Reading, arithmetic, logic, and language are mental artifacts, for their power lies in the rules and structures that they propose, in information structures rather than physical properties. Mental artifacts also include procedures and routines, such as mnemonics for remembering or methods for performing tasks. But whether physical or mental, both types of artifact are equally artificial: They would not exist without human invention. Indeed, anything invented by humans for the purpose of improving thought or action counts as an artifact, whether it has a physical presence and is constructed or manufactured, or whether it is mental and is taught.

The technology of artifacts is essential for the growth in human knowledge and mental capabilities. Think of where we might be without written history or without the development of logic, arithmetic, or skills of reasoning. Artists and musicians have often been the first to push new technologies, to explore their potential for enhancing our perceptions. But technology can enslave as well. It can be used as a drug, diverting people from more productive pursuits. Technology also gives power to those who have it, thereby disenfranchising those who do not.

I recently acted as a consultant to a company (which I am not allowed to name) that invited a group of people together to discuss an ambitious, large project (which I am not allowed to describe). Most of the people were noted film and television directors, producers, and writers. They told some of the methods they employ for inducing tension, horror, and excitement in their audiences. I was impressed.

I was entertained and enthralled by these people. Afterward, I eagerly watched some of their television shows and movies. These were bright, educated, intelligent individuals: I expected educated, intelligent television and films. Alas, the shows and movies were disappointing. What they demonstrated was yet another dumbing of the audience, yet another way to keep people trapped in their homes for hours a day, glued to the never-ending stream of entertainment on their video screens. In the case of the new technology I was being asked to give advice about, there was to be yet an even higher-technology means of entrapment.

Were the television shows and films clever? Yes. Engrossing? Sort of. Edifying? Enlightening? Informing? Absolutely not. Yes, if I watched the techniques, I could see the brilliance, but mostly aimed at keeping the action moving, at capturing and maintaining the viewer's attention. Yes, there were clever subtleties in the actions and settings, but mostly in the background and as asides to the main scenes and story lines.

The more I thought about the plans I had been asked to advise on, the more I thought them wrong. Not just wrong, but harmful. Yet another means of reducing the intelligence level of the watcher. Another way to make people willing victims of an increasing stream of commercials. But now high-technology commercials, ones that interacted with you, that got you to commit yourself and your credit card number even as you watched. Actually, you probably wouldn't even have to tell the system your credit card number: It would already know the number. It would know more about your likes and dislikes, personal activities, and bank accounts than you would realize, perhaps even more than you yourself would know. Was this the promised land of the new technology?

Howard Rheingold—writer, author, and editor of the *Whole Earth Review*—started a speech by telling us that he was an addict. He was hooked. He needed his daily dose, else his body rejected life and rebelled. Worse: He was also a pusher. He pushed his addiction on everyone he met, starting with his own family. Rheingold was a technology addict, a technology pusher. Television, computers, electronic games, household appliances. The electronic networks of the world brought thousands of mail messages and electronic bulletin boards daily to his computer screen. He communicated for hours a day with friends he had never met.

"But," Rheingold now tried to assure us, "I am aware of my addiction, aware of the dangers that it brings, the neglect of home and family that it causes." "Am I addicted?" I asked myself, listening to the talk. My home not only has computers, it has a computer network. I use electronic mail a lot. Perhaps it is actually using me: I get almost twenty thousand electronic mail messages a year. Rheingold was now reformed, but not yet free of the addiction. And like most newly reformed addicts, he proselytized many cures. He questioned the cause of his addiction, crusaded even, enlisting the old-fashioned technology of books against the new-fashioned technology of computers and communication.

I went out and bought the book Rheingold said had finally brought him to his senses, Jerry Mander's *In the Absence of the*

Sacred. I read the book and, just as Rheingold would have hoped, got depressed. Demoralized, but not entirely convinced. Mander thinks the negative side of technology is bad enough that much of it should be abolished. I think the positive side is good enough to keep, to cherish, to be expanded. I don't think we disagree about the potential benefits or perils. The difference is that he thinks that today the negative outweighs the positive. I remain more optimistic.

I maintain that much of our human intelligence results from our ability to construct artifacts. Yes, we humans have capable brains, but our brains are limited in power. But we, unique among all the animals, have learned how to overcome our limitations. We have invented tools that make us stronger than the unaided body would otherwise be, faster, and more comfortable. Our inventions warm us, clothe us, and feed us. Our technologies also make us smart: They educate and entertain us. Only the most rabid anti-technologist among us would want to do without clothing, shelter, and heat. Who would be ready to give up language, reading, and writing? What about arithmetic or simple tools? Why are some of the artifacts of technological development thought to be beneficial, others not?

Perhaps it is because technology has developed by accident. That is, technology has not been planned, it just happened. It started off slowly, with the use of simple tools in the wild. Modern humans made more complex tools than any other animal, more complex than Neanderthals or other hominids that preceded us. Whereas apes might trim a branch to make it into a probe for insects or use rocks to pound and break, modern humans learned to shape rocks into cutting instruments, then to tie the rocks to sticks to form spears. Eventually, modern humans learned to make clothing, to control fire, and from fire, they learned to extract metals from rocks and shape them into tools.

Although these early attempts to make tools and clothing were simple by modern standards, they were sufficiently complex to require a specialization of effort. The tools used for mining, metalworking, and carpentry are different from those for hunting, farming, cooking, and making clothes. Moreover, the skills and knowledge needed to use these tools efficiently are vastly different.

As a result, from the natural, unplanned development of tools came specialization in toolmaking. Those who were good at making one class of tools were distinguished from those who were good at making another class and, in turn, different from those skilled at using the tools. Those who were better at observing than hunting might stare for weeks and weeks at the sun, moon, and stars, apparently shirking their share of work, but also learning how to predict the seasons of the year, the good times to hunt, to plant, and to harvest. Each new advance of technology added to the powers and abilities of human society; each new advance also added to the amount of knowledge that newer generations would have to learn.

Each new discovery changed society to some extent. The background knowledge required more and more learning, thereby leading to more specialization. Cultural groups who had the most technology enjoyed some benefits over those who had less, so over the thousands of generations that human society has existed, power naturally drifted toward those with the most advanced technologies. Each technology added to the human ability to produce yet more technology; each required more and more skills to master, more and more schooling, more and more specialization. This means that society became divided into groups of those with some specialized knowledge and abilities and those without. But none of this was planned. The implications were not clear to the inventors, the developers, or the promoters.

I am not the first to have pondered the duality of technology.* But unlike many of those who have preceded me, my goal is to increase the general understanding of how these technologies of cognition interact with the human mind. Some critics of technology hold it responsible for much that ails modern society; they hold that technology inevitably leads to problems. I do not share these views. I believe that although technology has indeed contributed to many of today's problems, a perverse outcome is not inevitable. Technology, after all, has done much to benefit humankind. Some aspects of technology enhance life even as others diminish it. If we learn the reasons for and the properties of these various technologies, then maybe we can control the impact. So, yes, I am delivering a message of warning, but one accompanied by hope, not despair.

TOWARD A HUMAN-CENTERED VIEW OF TECHNOLOGY AND PEOPLE

Science Finds, Industry Applies, Man Conforms.
*Motto of the 1933 Chicago World's Fair**

In the past, technology had to worry about fitting people's bodies; today it must fit people's minds. This means that the old approaches will no longer work. The same analytical methods that work so well for mechanical things do not apply to people. Today much of science and engineering takes a machine-centered view of the design of machines and, for that matter, the understanding of people. As a result, the technology that is intended to aid human cognition and enjoyment more often interferes and confuses than aids and clarifies.

The evolution of science and engineering has moved toward more abstract, analytical analysis: Measurement and mathematical methods reign supreme. The bias toward a machine-centered approach is subconscious, which makes it ever more insidious: Those who follow the machine-centered approach are unaware that they are doing so, simultaneously denying the claim and defending their methods as logical, obvious, and necessary. Let me give some examples.

People, I have often been told, are easily distracted. Their attention wanders. In fact, people often complain about their own lack of concentration: "I get distracted by every new event in the room instead of concentrating upon my work." Distractibility is indeed a problem for those of us who must concentrate upon a task. It takes great mental effort to avoid distraction. This is clearly one deficit of the human mind.

Want another? What about the inability of many to speak grammatically? In the study of language and in the attempts to develop machines that understand human speech, one common complaint in the scientific research community is that most normal spoken speech is "ungrammatical," therefore posing great problems to the understanding. People seldom utter clean, precise sentences, the scientists point out. People's everyday speech is incredibly sloppy. People insert *ers* and *ums* into the sentences. They repeat words, start sentences over, and leave thoughts hanging, unfinished. Critical parts of the sentence are left out, and pronouns abound, not always well specified. Sometimes people select the wrong words, saying something that means just the opposite of what they intended. Worse, sometimes this is intentional: We say the opposite of what we mean in order to be ironic or sarcastic.

If distractibility and ungrammaticality are not enough, how about the insidious influences of emotion, clouding decision making and thought? How many times have you heard the complaint that someone has acted illogically, following emotions rather than reason? And what about human error? We all know that people err, sometimes with tragic results. What terrible creatures we are: distractible, ungrammatical, illogical, and errorful.

This view of the human, widespread though it may be, seems strange in the face of the evidence. If we are so inadequate, how did the human race manage to develop the technology that now causes us to be judged so poorly? The answer is that these characterizations are inappropriate.

A machine is not distractible: My computer will keep doing its job even as the building burns down. We criticize people for their distractibility, but do we really want undistractible people? Instead of complaining that people are distractible, we might rejoice in the fact that people are attentive to their surrounds and to new events. In other words, the same behavior that is a liability from the machine-centered point of view is a virtue when seen from a human-centered perspective.

What about the distractibility that prevents us from concentrating upon our tasks? Perhaps it is the task—and the insistence upon a strict deadline for its completion—that is the problem. Different tasks, or a different pace for the same tasks, might make the judgment of distractibility fade away. With the proper set of tasks, our continued attention to new events might become a virtue, not a deficit.

In similar ways, the statement that people speak ungrammatically is a peculiar value judgment. People speak the way people speak, and their utterances have been understood for tens of thousands of years. To say that people speak ungrammatically is to say that people don't speak the way our artificial grammars describe language. The same is true of rationality. To say that people often act illogically is to say that their behavior doesn't fit the artificial mathematics called "logic" and "decision theory." Just as the speech judged ungrammatical is perfectly understandable by its intended audience, the behavior judged illogical is understandable. It is simply based upon other considerations than go into the scientist's analysis. From a human-centered point of view, these are all reasonable, sensible ways of acting.

Today most industrial accidents are blamed on human error: 75 percent of airline accidents are attributed to "pilot error." When we see figures like these, it is a clear sign that something is wrong, but not with the human: What is wrong is the design of the technology that requires people to behave in machine-centered ways, ways for which people are not well suited.

How well is society doing at adapting technology to the minds of its users? Badly. We still suffer from the mind-set of the Chicago World's Fair: "Science Finds, Industry Applies, Man Conforms." We are overwhelmed with an onslaught of technological devices that have been designed from the machine-centered point of view, technological devices that confuse us, that alter normal social relations. Our self-created technological world controls and dominates us. The signs are clear, from confusion and difficulty in using household and office appliances to a heavy incidence of human error in industry. Yesterday a nationwide phone outage, today a computer failure that strands travelers and disrupts banking, tomorrow a major ship or airplane disaster: all attributed to human error.

When technology is not designed from a human-centered point of view, it doesn't reduce the incidence of human error nor minimize the impact when errors do occur. Yes, people do indeed err. Therefore the technology should be designed to take this well-known fact into account. Instead, the tendency is to blame the person who errs, even though the fault might lie with the technology, even though to err is indeed very human.

My goal is to develop a human-centered view of the technologies of cognition. My theme is not antitechnological, it is prohuman. Technology should be our friend in the creation of a better life; it should complement human abilities, aid those activities for which we are poorly suited, and enhance and help develop those for which we are ideally suited. That, to me, is a humanizing, appropriate use of technology.

But if we are to do all this, then we had better understand the way in which technology interacts with people and with human cognition. I am concerned that the mind has become the wasteland of modern technology, perhaps not unlike the ecological wastelands produced through the by-products of our industries. Ecological wastelands can result by accident, not by the intentional wish to destroy or pollute. But the process of making a desired product can also generate side effects and waste products that are of little value and sometimes toxic.

Likewise, within the ecology of the mind, we build our technologies to serve ourselves. In the past, these have primarily been physical technologies, the tools of food and comfort, power and speed. Today, more and more, we construct information technologies, the tools of entertainment, communication, and calculation. The ever-increasing collection of all imaginable statistics is meant to aid the decision maker and the corporate manufacturer. The increasing richness of the news and entertainment media is meant to inform and entrance. The increasing interconnection of the world into a vast communication network is meant to aid joint work and interaction. But the side effects can be chaotic.

Consider the erosion of personal privacy. In part, this comes about naturally, as unintended side effects of the introduction of cellular telephones, credit cards, and the computerization of banking records and store accounts. The telephone system always knows your exact location, and your credit card and other financial activities leave a precise record of both your location and your activities. These databases of personal statistics can be useful, for they allow one to place and receive telephone calls even while traveling and to purchase items almost anywhere in the world without the inconvenience of carrying large amounts of cash. But the mechanics of ensuring accuracy of information lag far behind the mechanics of collection and dissemination. The world may know myriad private details about individuals, but those details may be false, and once distributed, that false information may be impossible to eradicate.

We are in the midst of what some people call "the information explosion," but there is too much information for anyone to assimilate, the information is of doubtful quality, and perhaps most important, the things we collect statistics about are primarily those things that are easiest to identify and count or measure—which may have little or no connection with those factors of greatest importance. It is easy to collect statistics on number of hours worked, on cost of equipment, and on such statistical indexes as "productivity of labor." It is much more difficult to collect statistics on the quality of a product or its effect on the quality of life.

Around the start of the twentieth century, the science of time-and-motion studies developed powerful analytical tools for improving the work force by increasing its efficiency, decreasing its fatigue, and increasing the output of skilled workers.* The methods were—and still are—extremely effective at achieving their goals. The question is whether those goals were appropriate. Those efficiency experts thought they could make everyone happier: The worker would produce more, but with less physical and mental effort. The factory owner would get more output for less cost and could therefore afford both to pay the workers more and to charge less for the product. Everyone would be better off.

What the time-and-motion folks ignored was the person. They took a mechanistic, machine-centered point of view, analyzing every action through high-speed motion picture photography and doing controlled experiments that examined the differences among various procedures for the same tasks. They looked for wasted motion and inefficient procedures. They measured the speed of movement and determined the most efficient weights that could be moved and patterns of movements that could be made. Sometimes smaller weights or slower movements were superior to larger weights or faster movements, and all this was duly recorded and studied. Time-and-motion studies do lead to enhanced production in the short run. In the long run, they can lead to diminished quality of life, diminished quality of the product.

We humans are thinking, interpreting creatures. The mind tends to seek explanations, to interpret, to make suggestions. We are active, creative, social beings. We seek interaction with others. Unlike machines, we change our behavior as we attempt to understand what others expect of us. All of these natural tendencies are thwarted by the efforts of the engineering approach to efficiency. The machine-centered view is concerned primarily with operations per second. This approach emphasizes short-term productivity and treats workers in isolation from the social structure in which they participate. The result is a deterioration of long-term goals such as quality of product, satisfaction of the worker, and the need for a nurturing social environment.

Efficient, routine operations are fine for machines, not for humans. The body wears out—"repetitive stress syndrome," we call it today. Just as the body can wear out, so too can the mind—the syndrome called "burnout"—wearing out the ability to create, to innovate, or even simply to care about the work being produced. A worn-out mind leads to a demoralized worker, to someone who no longer cares about the job and who is apt to leave. Every time an employee leaves, the result is added cost and aggravation for the company: more costs to hire new people, costs for training, but costs that are not figured into the normal analysis of the engineer. Worse, the machine-centered approach to the design of a job leads to an uninspired society, where mental creativity is much reduced. None of this was planned: It is an accidental by-product of the age.

Accurate measurement underlies all of science. Without simple, reliable measurements, science would be denied the use of its most powerful tools: precise measurement, repeatable experiments under controlled conditions, and mathematical analyses. The problem is that when it comes to the human and social sciences, our measurement capabilities are limited at best.

Humans are extremely complex, the most complex entity ever studied. Each of our actions is the result of multiple interactions, of a lifetime of experiences and knowledge, and of subtle social relationships. The measurement tools of science try to strip away the complexities, studying a single variable at a time. But much of what is of value to human life results from the interaction of the parts: when we measure simple, single variables, we miss the point.

Call the sciences that rely on precise and accurate measurements the "hard" sciences. Call the sciences that must rely on observation and classification, on subjective measurement and evaluation, the "soft" sciences. And call the technologies that build upon and support these scientific opposites hard and soft technology. Now there is nothing wrong with hard science and technology. The problem is with what is left out. The hard side of science measures what it can and ignores the rest. The soft side tries to deal with those left-out parts, believing that to ignore them is to ignore critical and essential parts of the world.

The final result is that technology aids our thoughts and civilized lives, but it also provides a mind-set that artificially elevates some aspects of life and ignores others, not based upon their real importance but rather by the arbitrary condition of whether they can be measured scientifically and objectively by today's tools. Consequently, science and technology tend to deal solely with the products of their measurements, they divorce themselves from the real world. The danger is that things that cannot be measured play no role in scientific work and are judged to be of little importance. Science and technology do what they can do and ignore the rest. They are superb at what they do, but what is left out can be of equal or greater importance.

TWO KINDS OF COGNITION

Of all the accidental by-products of technology that help create the wasteland of the mind, the one I am becoming most concerned about arises through the technologies of entertainment, especially as they spill over to media of all forms, to education, and to the intellectual side of life. I am concerned that the new tools have moved us in unexpected ways to accept experience as a substitute for thought. I had better explain.

There are many modes of cognition, many different ways by which thinking takes place. The two modes particularly relevant to my analyses are called *experiential* cognition and *reflective* cognition. The experiential mode leads to a state in which we perceive and react to the events around us, efficiently and effortlessly. This is the mode of expert behavior, and it is a key component of efficient performance. The reflective mode is that of comparison and contrast, of thought, of decision making. This is the mode that leads to new ideas, novel responses. Both modes are essential for human performance, although each mode requires very different technological support. Without a good understanding of the differences between these modes coupled with an understanding of human perception and cognition, it is not possible to harness technology, to make its products appropriate for people.

It is dangerous to divide something as complex as human cognition into only two categories.* Human cognition is definitely a multidimensional activity, involving all of the senses, internal activities, and external structures. But focusing upon only two categories does indeed enable us to highlight and compare differing aspects of mental behavior. Let us concentrate upon only the two ways of being smart I have just mentioned: the experiential and reflective modes.

These two modes of cognition do not capture all of thought, nor are they completely independent: It is possible to have a mixture, enjoying the experiential mode while simultaneously reflecting upon it. Still, much of our use of technology seems to force us toward one extreme or the other. Some of the arguments about the value of television and other entertainment media stem from confusions between the relative nature of these two kinds of cognition. Many existing sets of instrumentation and equipment fail by providing reflective tools for experiential situations or experiential tools for reflective situations.

The solution is to ensure that we maintain the proper proportion of reflectiveness. We must understand its place and provide appropriate technological support and intellectual training. Modern technology has the power to enhance reflection, to make it ever more powerful than before. While the experiential mode of cognition can be practiced simply by experiencing it, reflection is more difficult. Anyone can reflect, and probably everyone does: It is a natural, human state. But effective reflection requires some structure and organization. Reflection is greatly aided by systematic procedures and methods, and these are learned primarily by being taught. Alas, our educational system is more and more trapped in an experiential mode: the brilliant inspired lecturer, the prevalence of prepackaged films and videos to engage the student, the textbooks that follow a predetermined sequence. We strive to keep our students engaged in our schools by entertaining them. This is not the road toward reflection.

Mind you, I am not trying to eliminate experiential cognition. I enjoy the experiences as well as anyone else. A valuable part of life is the ability to immerse oneself in the happenings of the world, whether it be the thrill of the water and wind while sailing or the thrill of the activities imagined while reading a well-crafted book. The mode is experienced most fully when the person is actually participating, actively engaged in doing the activities, but it can also be appreciated vicariously, watching or reading or imagining. Moreover, experiential thought is essential to skilled performance: It comes rapidly, effortlessly, without the need for planning or problem solving.

The skill of an expert is that of experiential cognition. The skilled sports player must be in experiential mode, responding automatically to the events of the game. So too the experienced pilot or even the mathematician: Look, see, respond. Certainly, if the airplane in which I were traveling had a problem, I would not want a pilot who had to reflect on what to do: I would want a pilot who was so well trained that the proper responses would come effortlessly, quickly. Experientially.

But the enjoyment of experiential mode is also its danger. It seduces the participant into confusing action for thought. One can have new experiences in this manner, but not new ideas, new concepts, advances in human understanding: For these, we need the effort of reflection. But the worse problem is probably that of vicarious experience, when one uses the technology of film, video, or even the printed page to watch others in experiential mode. Vicarious experiences can be entertaining, but they cannot substitute for active participation.

TWO

EXPERIENCING THE WORLD

IN THE MUSEUM OF SCIENCE AND INDUSTRY IN CHICAGO THERE IS AN exhibit donated by General Motors Corporation that tells how science is used in automobile design and manufacturing. The exhibit is a movie/video presentation. There is a large screen, and as you stand in front and watch, the image of a scientist dressed in a white laboratory coat comes on screen. The scientist explains the marvels of science, sings, dances, jokes, and juggles. The display itself is in a limited form of three-dimensional projection, with images appearing at several different depths. The objects that the scientist describes sometimes appear in space in front of him, sometimes behind him. Some images appear to be video projections, others appear to be objects floating majestically in space, neither real objects nor images projected onto a screen.

I visited the exhibit when the Cognitive Science Society held a banquet there during one of its annual conferences. I was accompanied by a number of colleagues. We are real scientists, as opposed to the one shown in the display. We don't wear white lab coats when we work, nor can we sing, dance, or juggle (at least not well enough that anyone else would want to listen or watch). We noted that the "scientist" in the exhibit didn't actually say anything that had content. We listened to the words and waited patiently for knowledge: None ever came.

The most intriguing part of the exhibit was the exhibit itself. How did it work? How could the images appear at different depths? Careful observation indicated that there were only a few fixed locations where images could appear. Even so, how was this done? Were there several internal screens, with different images projected on them at different times? And what about those "virtual images"? If there had been only one, it would have been easy for us to figure out how it might have been projected, but there were quite a number of them. Were they from real objects on a rotating turntable, perhaps rotated and illuminated at the proper time, the images then projected by lenses or mirrors?

We learned a lot from the exhibit, but not in the way that the museum had intended. The exhibit didn't aid us, nor did it teach anything about science—it was an advertisement for General Motors. No science at all, except inside the display itself, but because that was not the point of the exhibit, no effort was made to capitalize on that most interesting and informative part of the display. The museum attendee was left with pure entertainment and no substance.

The problems with the exhibit are similar to those found in many science museums. I once talked to a director of a major science museum who admitted this but said that the real problem was to get people excited about science in the first place. "Museum visitors," the director patiently explained to me, "don't want to read lengthy descriptions or to hear the details of science. We have done our job if we can get them excited by the phenomena of science."

The museum director may have a valid point, but the exhibit at the Museum of Science and Industry illustrates the contradiction within the argument: The expensive, fancy exhibit did indeed entrance many visitors, but it did not leave them with any new knowledge. Wouldn't it have been wonderful if we could have peeked in the rear of the exhibit, if the museum had exploited this opportunity to teach something about optics and vision?

My students and I have spent many hours observing and talking with visitors at a number of science museums. People really do enjoy them. They appear to be having fun, they tell us they like them, and they intend to come back. But do they understand what they have experienced? Not necessarily. Have they learned anything in the process? Surprisingly little. Worse, on occasion, people would explain to us what they had learned, but the explanations were wrong: False information seemed almost as readily transmitted by the exhibits as true information.

There are exceptions to the story, the one most familiar to me being the Exploratorium in San Francisco, a science museum that lacks the glamour and professional polish of typical museums but instead emphasizes active encounters on the part of the museum visitor. The Exploratorium has a large team of eager recruits (called "explainers") who are trained to help visitors, to guide them through the steps of the exhibits, and to answer questions, something few other museums provide.*

Note that my complaints are shared by others.* The psychologist Richard Gregory (who helps run the Exploratory science museum in Bristol, England) comments that "looking at the traditional museums of science we find remarkably little science. It is hard to find Kepler's or Newton's laws; how spectral lines may be related to atomic structure; concepts of Quantum Physics or relativity. It is quite hard to find clear explanations of how motors, or radios or freezers work."

The science museums have noted the criticism and have risen to the defense of the emphasis on experiential cognition: "Is the goal of motivation—a prerequisite and prime constituent for the public understanding of science—reason enough for funding interactive science?" responded Melanie Quin, director of the European Consortium for Science, Industry and Technology Exhibitions. "Yes, of course," she continued. "The goal of raising enthusiasm and opening the door to a new interest is justification for supporting an art gallery. . . . So why not a modern centre for science events?" The point is well taken. Experiential learning isn't enough, of course, but it is indeed a good motivator and therefore a sensible starting point. As Quin puts it: "there is a general consensus . . . that exhibits on their own are not good at teaching. They are about inspiration. And interest, once aroused, must be taken advantage of—by schools, and through all the activities other than exhibits developed by museums and science centres." How could I not agree with such an eloquent statement? Nonetheless, I worry that too many museums use such soothing words as excuses for the lack of follow-through with the hard work of providing opportunities to learn, to think, to reflect.

An interesting example of how one can combine experiential cognition as a motivator with tools for reflective learning is provided by video arcade games. In the video game business, the game designers provide an attractor mode, an experiential display that shows the passerby the major features of the game and tries to entice the person to play. Once interacting with the game itself, the player has to use both experiential and reflective cognition to be effective: reflective mode to learn the secrets and develop the best strategy, experiential mode to enjoy the situation and also to be at the most appropriate skill level of responding. Actually, the attractor mode has its reflective aspects. The skilled player studies the scenarios displayed in the attractor mode, for they often reveal hints and secrets regarding appropriate ways to respond to particular situations that will be encountered within the game. Game manufacturers have figured out both how to attract people to the game and how to instruct them appropriately. Why can't the museum developers be as clever?

EXPERIENTIAL AND REFLECTIVE COGNITION

The essence of expertise is knowing what to do, rapidly and efficiently. The pilot pushes the throttles forward, controlling the nosewheel and rudder to keep the plane on the runway. "Rotate," says the copilot, and the pilot pulls back on the control wheel and, as the airplane climbs, adjusts the flaps, throttles, and gracefully banks the plane. All this is done with practiced ease and skill, continually integrating numerous sources of information—the scenes out the window, the spoken comments of the copilot, the readings of the flight director and engine instruments, the felt locations of the throttle and flaps, the sounds of the engines, and the previously memorized procedures for the particular airport and runway.

Similar stories can be told of any expert performance, whether in the airplane, on the playing field, or in front of an audience.

Consider how you can be captured by an exciting novel or television show—so engrossed that the rest of the world temporarily fades from existence. These are examples of experiential cognition: The patterns of information are perceived and assimilated and the appropriate responses generated without apparent effort or delay. Experiential thought is essential to skilled behavior. It appears to flow naturally, but years of experience or training may be required to make it possible. Although I call the mode "experiential" to emphasize the subjective aspect, an equally valid name for this mode of cognition would be "reactive," emphasizing the automatic nature of the reactions.

All of us are expert in some domains, not in others. Once we are expert, the required responses appear to come effortlessly. A tremendous amount of processing is still required, but it is all done without conscious awareness. Subconscious processes match current experiences with our huge storehouse of experiencc and knowledge. Decisions that require considerable insight and information can thereby be made rapidly and without apparent effort. Experiential thought is reactive, automatic thought, driven by the patterns of information arriving at our senses, but dependent upon a large reservoir of experience.

Try it yourself: What is the sum of 2 and 4? The answer comes without conscious reflection, without any awareness. So it is with experiential cognition: The responses occur automatically, without any need for reflection, but as in this case, only if the required information has already been acquired, and this can often take considerable time and effort.

Reflective thought is very different from experiential thought, even when both are applied by the same people in similar situations. To see this, consider once again our pilots in their cockpit. We have already noted that routine, well-practiced flying is primarily experiential in nature: Encounter the situation, make the appropriate response. But suppose some decision making is required. Suppose the pilots have to plan. Now the situation calls for reflection. Thus suppose the airplane radar shows a large storm on the flight path, perhaps impeding the plane's ability to land at the scheduled destination. The crew has to decide how best to avoid the storm. They could try to go above or detour around. Either choice is problematic, for the exact location of the storm is not known. Moreover, each will prolong the flight, use more fuel, and possibly make it impossible to get to the scheduled destination. The flight crew has to compute the likelihood of getting through to their destination and the estimated flying time and fuel usage for each possible alternative. Any alternative routing has to be cleared through the air traffic control system and, if a new destination is required, communicated to the airline company. This task involves considerable planning. The flight crew will discuss the alternatives with one another and perhaps do some numerical computations. The various alternatives will be compared until finally a decision is made.

The difference between the two modes is rooted in the technical details of the information-processing structures of the brain. The one mode, experiential, involves data-driven processing: Something happens in the world, and the scene is transmitted through our sense organs to the appropriate centers of mental processing. But in experiential mode, the processing has to be reactive, somewhat analogous to the knee-jerk reflex. You know, tap the knee in just the right spot on the patellar tendon with a hammer and the lower leg jerks upward. In the case of the reflex, the stimulation of the hammer tap goes up to the spinal cord, makes the connection with the nerve fibers controlling the leg muscles, and zap, the leg jerks. No thought is involved. Experiential processing does involve some thought, but it is similar to the reflex in that the relevant information must already exist in our memory and the experience simply reactivates that information, much as the hammer tap activates the muscle movements.

Actually, some simple deductions are possible. Computer scientists Lokendra Shastri and Venkat Ajjanagadde have made an important first attempt to show just how much deduction is possible with experiential processing (which they call reflexive reasoning). They show that there are strict limits to how far a chain of reasoning can proceed. But, argue Shastri and Ajjanagadde, reflexive reasoning can make simple deductions. Thus if you read or are told that cats attempt to attack birds and if you already know that Sylvester is a cat and Tweety a bird, you can rapidly infer both that

Sylvester is likely to attack Tweety and that, as a result, Tweety is frightened of Sylvester, all reflexively, experientially. This type of rapid inference is essential to explain the ease with which we read books. After all, in most simple reading, we must make these kinds of inferences to understand the meaning, yet we do not continually stop and ponder each sentence as we read it, certainly not with everyday, nontechnical material.

Reflective reasoning does not have the same kind of limits on the depth of reasoning that apply to experiential cognition, but the price one pays is that the process is slow and laborious. Reflective thought requires the ability to store temporary results, to make inferences from stored knowledge, and to follow chains of reasoning backward and forward, sometimes backtracking when a promising line of thought proves to be unfruitful. This process takes time. Deep, substantive reflection therefore requires periods of quiet, of minimal distraction. Moreover, the use of external aids facilitates the reflective process by acting as external memory storage, allowing deeper chains of reasoning over longer periods of time than possible without the aids.

It would be wrong to try to determine which mode of cognition is superior, experiential or reflective. Both modes are needed, and neither is superior to the other—they simply differ in requirements and functions.

The reflective mode is that of concepts, of planning and reconsideration. It is slow and laborious. Reflective cognition tends to require both the aid of external support—writing, books, computational tools—and the aid of other people. The external representations have to be tuned to the task at hand if they are to be maximally supportive of cognition. Reflection is best done in a quiet environment, devoid of material save that relevant to the task. Rich, dynamic, continually present environments can interfere with reflection: These environments lead one toward the experiential mode, driving the cognition by the perceptions of event-driven processing, thereby not leaving sufficient mental resources for the concentration required for reflection. In the terms of cognitive science, reflective cognition is conceptually driven, top-down processing.

The experiential mode of performance is one of perceptual processing: what cognitive science calls pattern-driven or event-driven activity. The human perceptual system is well suited for the experiential mode, hence our excellent abilities at sports and other physical activities, our expert driving and piloting of aircraft. Experiential mode plays important roles in the routine aspects of otherwise reflective tasks, such as in some phases of chess games where the perceptual recognition of the game state can lead to a well-learned, pattern-driven response without deep reflection or planning.

From the scientific point of view, thought is a complex activity that involves multiple operations and components. It is important to remember that the dichotomy I have presented of two distinct modes is somewhat simplified. The two do not capture all of thought, nor are they completely independent of each other: It is possible to have a mixture, enjoying the experiential mode while simultaneously reflecting upon it. Most cognition involves components of both. Some kinds of cognition—daydreaming, for example—are difficult to classify as either.

From a practical point of view, the distinction between experiential and reflective thought is worth considering, in part because much of our technology seems to force us toward one extreme or the other. With proper artifacts, we can enhance each mode.

Tools for experiential cognition should make available a wide range of sensory stimulation, with enough information provided to minimize the need for logical deduction. Similarly, tools for reflection must support the exploration of ideas. They must make it easy to compare and evaluate, to explore alternatives. They should not restrict behavior to the experiential mode. In both cases, reflective and experiential, the tools must be invisible: They must not get in the way. If tools are designed inappropriately, or for that matter, if appropriate tools are used in inappropriate ways and places, various dangers may arise:

- *Tools for experiential mode behavior that require reflection:* These tools turn simple tasks into problem-solving exercises, causing needless mental effort, taking needless time. When taking pictures with a camera or driving an automobile, it is essential to be able to react quickly and effortlessly. If the camera or automobile controls require reflection, performance suffers: Consider the lapses of attention from driving when attempting to change the station with the typical automobile radio.

- *Tools for reflection that do not support comparisons, exploration, and problem solving:* In many cases, we need to be able to look over the situation and compare alternative courses of action or perhaps just ponder and reflect upon the variables involved. The most common tools for this purpose are writing and drawing. Many electronic decision aids tend to restrict the availability of information to small segments visible on the relatively limited display. This makes it difficult to integrate disparate sources of information, difficult to explore and to make comparisons.
- *Experiencing when one should be reflecting:* The experiential mode leads to responses without thought, without contemplation. This is essential when events move rapidly, but if the situation changes, experiential cognition may not be flexible enough to change appropriately.
- *Reflecting when one should be experiencing:* Too much reflection and the world will pass one by. See every point of view, consider every possible alternative. See the merits and perils of each alternative. Flutter in the breeze of public opinion, get caught in the trap of thought, never to decide and act.

Of all these dangers, the one I think poses the greatest peril today is that of experiencing when one should be reflecting. Here is where entertainment takes precedence over thought. Worse, one can believe that the experiential mode has substituted for independent, constructive thought, for reason and reflection.

Reflective thought is the critical component of modern civilization: It is where new ideas come from. Trashy novels are more popular than serious, philosophical ones. Comic books more popular than novels. Films of fantasy and horror are more popular than films of content. Even informative news programs—documentary and discussion shows—are framed in the experiential mode, never allowing time for reflection, never allowing time for viewers to have their own thoughts. Unfilled time on the broadcast stations is thought to be nonproductive time. Why, the viewers might have their own ideas! Horrid thought. Worse, the viewers might get bored and do something else.

All work and no play makes for a nonrewarding life. All play and no work does not promise well for human survival and advancement. In the mental world, the correlates of entertainment and work are the two forms of cognition: experiential and reflective. Just as both entertainment and work are essential for a full physical life, both experiential and reflective thought are essential for a full mental life. The difficult intellectual challenge for modern society is to strike the correct balance.

TWO KINDS OF COGNITION, THREE KINDS OF LEARNING

Just as there are several kinds of cognition, there are several kinds of learning. Some time ago, my colleague David Rumelhart and I suggested that there were at least three different kinds of learning: accretion, tuning, and restructuring.

Accretion

Accretion is the accumulation of facts. This is how we add to our knowledge, learn new vocabulary or perhaps the spelling of an already known word. Suppose you chat with a friend or a neighbor and learn about current events and trade gossip: This is learning by accretion, adding to the stockpile of knowledge. When you already have the proper conceptual framework, accretion is easy, painless, efficient. Little or no conscious effort is required under these circumstances. However, when there isn't a good conceptual background, then accretion is slow and arduous. In this case, it can be difficult to learn the material. It requires repeating the material over and over again (rehearsing), using mnemonic strategies, or writing down the information.

Tuning

In between the initial stages of novice performance and the skilled, smooth performance of the expert are hours and hours of practice. Teach children the rules of arithmetic and they still cannot do even simple operations such as single-digit subtraction or multiplication with ease. Yes, they may always get the correct answer, but only after prolonged, laborious effort. The same holds for many motor skills, such as typing, playing musical instruments, and sports.

What does practice do? It tunes the skill, shaping the knowledge structures in thousands of little ways so that the skill that in the early stages required conscious, reflective thought can now be carried out automatically, in a subconscious, experiential mode. Experiential thought is tuned thought.

Tuning is a slow process. I once estimated that, in virtually any complex activity, it takes a minimum of five thousand hours to turn a novice into an expert. That is about two years of full-time effort. In fact, five thousand hours isn't really enough. You certainly would not be an expert violinist, skier, or tennis player in only two years. You wouldn't have reached true experiential modes of behavior.

Note the discrepancy between the apparent ease and automatic nature of the expert's experiential behavior and the laborious effort over a period of years required to reach that stage. Moreover, expert behavior must constantly be retuned. If the expert does not practice or otherwise use a skill, performance deteriorates. This is true of both motor and intellectual skills. Tuning is necessary to reach expert levels of performance, and then essential to maintain them.

We delude ourselves if we believe that skilled behavior is easy, that it can come about without effort. We forget the years of tuning, of learning and practice it takes to be skilled at even the most fundamental of human activities: eating, walking, talking, reading, and writing. It is tempting to want instant gratification—immediate expert performance and experiential pleasure—but the truth is that this primarily occurs only after considerable amounts of accretion and tuning.

Restructuring

The difficult part of learning is forming the right conceptual structure. Accretion and tuning are primarily experiential modes. Restructuring is reflective. Restructuring is the hard part of learning, where new conceptual skills are acquired. And the trick in teaching is to entice and motivate the students into excitement and interest in the topic, and then to give them the proper tools to reflect; to explore, compare, and integrate; to form the proper conceptual structures. Experiential mode is fine for the "attractor mode" of learning, fine for accretion and tuning. Reflective mode is essential for restructuring. The problem is to make the students want to do the hard work that is necessary for reflection.

Many years ago, I conducted a number of studies on learning. I was trying to increase my understanding of alternate ways of presenting material to groups of students and to discover when it was best for the instructor to intervene. My careful studies made only a relatively small impact on the rate of learning compared to the huge differences between students who seemed to be more interested in information and those who were less interested. Motivation turned out to be far more powerful than the cognitive variables I was addressing. Those students who were highly motivated learned the material far better than those who were disinterested. At the time, I gave up, for I had no idea how I could study motivation.

The world of entertainment deals almost entirely with motivation. The goal is to get people engaged in watching the entertainment regardless of their original interests. Talk to a filmmaker or a television producer and you soon learn numerous tricks of the trade for making material appealing and compelling.

Entertainment exploits the experiential mode, and even though well-crafted entertainment also requires reflection to be fully appreciated, this is not its major function. Nonetheless, entertainment can provide the impetus for reflection. Once people are curious about the questions, then they are stimulated and willing to do the work involved in pursuing the answers. Many people read the sales literature on major purchases *after* they have made their purchase. Once you have bought the new automobile or kitchen appliance, then you become more motivated to find out about it in great detail, perhaps even to read the sales literature on competing items to remind yourself of why your choice is superior. I have noticed the same with travel. After I have visited a foreign country, then I am much more motivated to study its history and current political status. Yes, it would make much more sense to have done the reading before making the purchase or visiting the country. Indeed, I usually go out and collect relevant books and magazine articles first, but they tend to remain unread until after the trip. Reflection is hard work, after all, but it can be pleasant when there is a reason for the effort.

OPTIMAL FLOW

Motivated activity, whether experiential or reflective, can be challenging and rewarding. The mind is captured, the experience is exhilarating. It is what experimental psychologist Mihaly Csikszentmihalyi calls a peak experience, an optimal flow. If you want a sustained, optimal experience, says Csikszentmihalyi—whether at work, home, or play—the important thing is a continual flow of focused concentration: absolute absorption in an activity.

Probably all of us have experienced this engaged state of focused attention, a form of trance. As all attention is focused upon the task at hand, the outside world fades away: Its noises and distractions subside. This trance world can be induced by many things, by books, plays, television. By games or music. By concentrated experiential cognition or by intense, focused reflection upon a problem. It is an enjoyable state, for when attention becomes so intensely focused upon the thing of interest everyday worries and fears are transcended and all else recedes. One lives for that task alone.

This focused concentration is easiest to sustain when in an experiential mode, when the experience is driven by the events. The good entertainer strives to establish this flow state, forcing the audience to surrender to the activity. The same state can also be reached during reflection, the major difference being that the state is now self-controlled, no longer dependent upon continual external stimulation. The trick in both cases is to avoid interruption. Whether the interruption comes from an outside source such as the ringing telephone or is self-induced through wandering, daydreaming thoughts, the end result is the same: disruption of the focused attention, disruption of the trancelike state of concentration.

Is there some way of achieving this state of optimal flow while learning? Note that people are typically willing to exert great mental effort upon their recreational but not their educational activities. Yes, one is done for enjoyment, the other assigned as tasks or duties in the schoolroom or on the job, but the difference seems paradoxical, especially since many people will tell you that the educational work is more important. The difference appears especially paradoxical if you simply compare what has to be done for each: The activities for recreation and education are essentially identical.

Think of what it takes to learn a game compared to what has to be done in school. To play a game well requires the same kinds of learning, study, understanding, and practice as are required of any educational activity. There is no reason why the learning and studying required in education should not be as captivating and enjoyable as the learning and studying of the game.

It is remarkable how little scientific knowledge we have about the factors that underlie motivation, enjoyment, and satisfaction. The issues seldom arise within the context of laboratory studies of human cognition. In part, this is because the logical, systematic, disembodied intelligence of the controlled studies leaves out subjective feelings, emotions, and friendly social interactions. That's the tradeoff between hard science, which requires things to be measured with precision, and soft science, which attempts to study those things for which measurement is difficult or impossible. As a result, we know little about how best to structure tasks and events so as to establish, maintain, and enhance the experience. Much of what we do know comes from the work of psychologists such as Csikszentmihalyi.

"Studies of flow," says Csikszentmihalyi, "have demonstrated repeatedly that more than anything else, the quality of life depends upon two factors: how we experience work and our relation with other people." And what factors are involved in that experience? To a large extent, activities that support a positive flow experience are those that "have built-in goals, feedback, rules and challenges, all of which encourage one to become involved in one's work, to concentrate and lose oneself in it." Leisure time does not have these properties unless individuals establish for themselves leisure activities that provide them. Work and school can provide settings for them but seldom do.

Two sets of studies shed some light on what it takes to invoke this optimal flow: Brenda Laurel's analysis of "first-person" experience and Susanne Bødker's "human activity approach."* Both emphasize the subjective feeling of the person; both discuss ways to minimize disruption.

Laurel started her studies by examining two activities that are extremely successful in capturing people's attention: games and theater. She distinguishes between two modes of participation: "first-person" and "third-person." Third-person engagement is passive, watching as a spectator. As we know from the plays and films that have successfully engaged us, this can be an effective way to participate, but the subjective experience is that of an outsider looking in, detached from the events. A first-person experience occurs when the person is directly and emotionally involved in the activities. This can happen in theater, movies, or sports if you subjectively project yourself into the event, sharing the suffering and disappointments as well as the rewards and successes.

It is much easier to have this experience when there are no distractions to interrupt. In the theater or stadium, the distractions come from unruly audiences or other events not related to the main activity. In the home or school, the distractions come from other people or the uncontrolled ring of the telephone. It also helps if the perceptual experience is optimal. The further away your seats are from the events in a theater or stadium, the more difficult it is to become involved. Whether at home or in a theater, a large screen with high-quality sound improves the ability to be captured by the event. The relatively small screen and sound system of the typical home television set distance the viewer from the event. As a result, the events occurring within the home compete with what is happening on the television screen. Replace the television set with a wall-size display, add surround sound, and the experience changes:

Now the television can create a full experiential state. Another way to reduce distractions and increase concentration is to wear a headset. In any environment the event best captures the attention when the sensory experience is maximized and distractions are minimized.

As Laurel points out, who knows better how to sustain interaction and interest than those who create theatrical experiences? There is a vast body of experience and theory to explain how best to generate and maintain the appropriate attentional flow and sense of participation. Once sensitized to the notion of activity flow, it soon becomes apparent why so many situations prevent the optimal flow state from forming or, once formed, from lasting.

Mind you, interruptions need not come from an external source: Sometimes they are caused by the tool being used for the task. This is especially true of the computer, with its wide range of notices, announcements, messages, and "dialog boxes." Many are unnecessary, and their presence serves to interrupt the train of concentration so essential to proper conduct of the task.

Susanne Bødker has studied these kinds of disruptions, pointing out that in performing a task, the person has a focus and a goal. All attention should be concentrated upon the task itself, not upon the tool. When the tool calls attention to itself, that creates a breakdown in the work flow. Tools should stay in the background, becoming a natural part of the task. What is needed is "direct engagement," the feeling of directly working on the task. With properly designed tools, the experts use them subconsciously, automatically. The tools, the person, and the task meld into a seamless whole.

We still have a great deal to learn about the nature of intense, rewarding experiences, whether at work or at play. But the current findings do suggest some guidelines. I offer them here, with the caveat that they have not all been sufficiently studied or analyzed. As the standard saying goes: More work needs to be done. Based on what we know today, the environment conducive to optimal experience should:

- Provide a high intensity of interaction and feedback
- Have specific goals and established procedures

- Motivate
- Provide a continual feeling of challenge, one that is neither so difficult as to create a sense of hopelessness and frustration nor so easy as to produce boredom
- Provide a sense of direct engagement, producing the feeling of directly experiencing the environment, directly working on the task
- Provide appropriate tools that fit the user and task so well that they aid and do not distract
- Avoid distractions and disruptions that intervene and destroy the subjective experience

Games, especially action games, are stimulating and compelling because they are event-driven activities, always presenting some new challenge to the player, maintaining attention by continual new stimulation, new challenges. This is one of the powers of experiential mode: The mind is externally driven, captured by the constant arrival of a barrage of sensory information.

Most education today is still delivered in the classroom with hour-long lectures by instructors. Students, however, cannot keep their attention on a topic for an hour: Nobody can. Csikszentmihalyi has done studies of what students "taught by an excellent teacher at a great high school near Chicago" think about in the classroom. The answer is surprising—almost anything *except* what the teacher is talking about.

> Using a pager on the teacher's desk, we compared what the teacher was saying when the pager went off, to what the students were thinking at the same time. The teacher was lecturing about how the troops of Genghis Khan were coming down in the west of China, outflanking their adversaries along the Great Wall, then moving north to conquer Yenching. What were the students thinking about? Of the 27 students in the class, only two were thinking about anything remotely resembling China. Most were thinking about the lunch they were expecting, the weekend they were looking forward to, a boyfriend or girl-

> friend, or some sporting event. Of the two students who were thinking about China, one was recalling a meal his family had at a Chinese restaurant the previous week; the other was wondering why Chinese men wore pony tails.

People are not good at keeping to a single task for long durations. William James, the great psychologist-philosopher who lived and worked at the beginning of the twentieth century, once estimated that his own attention span was approximately ten seconds. If this be so, then the ordinary classroom provides just the wrong situation, for the group setting demands concentration by the student with no means of maintaining attention. Studying is usually thought of—and taught as—a solitary activity. There is no reason for it to be solitary, no reason not to have numerous external aids, as long as the other people and the external information contribute to keeping attention focused.

There is an important difference between playing and practicing, doing an activity and learning that activity. Just doing something does not necessarily lead to learning. This point is well understood in sports instruction. Coaches distinguish between unsupervised play and training. You could play for a hundred hours and learn less than from a half hour of properly supervised training. In training, the coach carefully sets up the condition to be practiced and provides appropriate feedback and guidance so that the player can benefit from the experience. The same thing is true whether the practice be of a sport, chess, or mathematical recreations. In nonschool activities, the person often does the practice and training voluntarily, despite the fact that they require mental effort, considerable time, and dedication.

To be a good player requires concentration and focus: the essence of experiential mode. But to learn, to improve, to train oneself—that requires reflection upon the performance, the better to know what to change and what to keep. A coach provides the reflection for the player. Without a coach, self-reflection is required to think about and analyze one's own activities. Self-reflection is more difficult than the reflections provided by the coach. The coach, moreover, knows the proper mix of accretion, tuning, and restructuring and can assign practice exercises accordingly.

Similar processes can take place informally. Just go to the local video game arcade and watch the intensity and concentration of the players. Many are engaged in systematic exploration and training: coaching one another; passing on hints and shortcuts; learning, not just playing. The arcade, of course, provides continual stimulation, challenge, and reward. Once players have been captured by the simulated world of the arcade game, their attention and motivation take care of themselves.

Maybe this is the environment we need for education, whether in school, at work, or by oneself at home: continual stimulation, simulated worlds, and the proper social interaction with other players and teachers to ensure that there is guidance and feedback, so that the activity is a true learning, coaching, training activity—in short, so that it is educational. The new world of high-technology, multimedia education captures the experiential nature of the optimal experience, but at the neglect of the rest. It should be possible to offer both experiential and reflective experiences, all within the envelope of a sustained, optimal flow. Multimedia, the combination of television, computers, textbooks, and audio. Ah, the words that are used to describe it: *engaging, entrapping, creative, captivating, flowing*. Finally, a technology that will overcome the poor education provided by our school systems. Or will it?

As an educator, I look upon the work in multimedia with mixed feelings—a combination of enthusiasm and dismay. Of course, as an educator, I am tainted by all that is wrong with the school system. Nonetheless, let's look more carefully at what we have.

Multimedia

The Good	The Bad
The flow experience	The cuteness, the clever transitions and musical phrases. The entertainment aspects of multimedia.
The excitement that is felt	The wasted opportunities. The medium seems to be better at making time pass effortlessly than at making any attempt to exercise the mind.

The commitment by the users	The lack of depth. The apparent fear of boring the users by staying on one topic for too long, thus precluding any depth.
The technological sophistication	The technology: slow, plodding, interrupting the pace.

A recent review of education described multimedia the following way:

> Recent developments in computerized interactive multimedia can take us considerably further. . . . A trove of information lies just beneath the surface. . . . A photograph of some rather primitive-looking safety pins, for example, yields this caption: "American ingenuity: Inventor Walter Hunt created the safety pin in three hours. He patented it in 1849, and later sold all rights to the pin for $400." *(Leonard, 1992)*

This was meant to be a favorable review, but think about that caption: "Inventor Walter Hunt created the safety pin in three hours. He patented it in 1849, and later sold all rights to the pin for $400." This is depth? This is taking us "considerably further"? Than what? What have I learned besides a smattering of insignificant information: trivia amplified through colorful computer graphics, but trivia nonetheless.

When I watch children playing video games at home or in the arcades, I am impressed with the energy and enthusiasm they devote to the task. Mind you, these are not simple games. They can take days or weeks to play; they require a large amount of knowledge, exploration, and hypothesis testing. They require problem solving—saving the current state of the game and tentatively exploring novel states, then comparing the results, returning to the saved state when necessary. They require study and debate among fellow players and the reading of hint books. They require reflection. In other words, the games require just the behavior we wish these same children would apply to schoolwork. What is the difference between these informal experiences and the formal, structured behavior of the classroom? There is something captivating about informal learning. Why can't we get the same devotion to school lessons as people naturally apply to the things that interest them?

Educators can bore. Going back to the traditional classroom is not the answer. The fifty-minute class is not the answer. Ordinary people—you and I—cannot concentrate on a single topic for that long. Lectures are not the answer, no matter how good the lecturer. In contrast, there is much to be learned from the game makers. They obviously know how to capture interest sufficiently well that real learning takes place, albeit learning of irrelevant subjects.

The solution is to merge what each group of people can do best. Educators know what needs to be learned; they are simply pretty bad at figuring out how to get the intense, devoted concentration required for the learning to take place. The field of entertainment knows how to create interest and excitement. It can manipulate the information and images. But it doesn't know what to teach. Perhaps we could merge these skills. The trick is to marry the entertainment world's skills of presentation and of capturing the users' engagement with the educator's skills of reflective, in-depth analysis.

Science museums have grown up around the world. Museums of technology, with fancy, sexy exhibits. Push a button, watch something move. See this phenomenon and that. But what is learned? I have watched people in these museums and talked with curators. If the people stick with an exhibit for as long as two or three minutes, the curators are delighted. Two or three minutes? How on earth can anyone ever learn anything in two or three minutes? It takes hours. Recall the estimate of five thousand hours to turn a novice into an expert (and even this isn't really enough). Granted, we don't expect the science museum to produce experts, but two or three minutes?

There are some really excellent exhibits in some science museums where one can see and explore phenomena impossible to see anywhere else. Explore a coal mine, watch an operating beehive, experiment with reflection and refraction of light beams. The best of exhibits provide interactive, exploratory laboratories in which the phenomena of textbooks are exposed for study. In fact, many scientists say they have seen the best illustrations of their own field of study in some of these museums. Perhaps this is the model that we should be following. Perhaps we should use the technology to provide a rich database of information and demonstrations. Provide a learning laboratory in which students explore and solve problems that are posed by their teachers. In this way, teachers become assistants in the discovery of knowledge, guides to the exploration, reflection, and restructuring of the student's understanding. This is better for student and teacher alike.

What are the differences between the formal, structured behavior of the school and informal experiences such as the video arcade or practice at sports or, for that matter, learning the material of one's hobby or recreation? The following table lists some of the basic differences:

Informal Learning	**School Learning**
Unstructured.	Structured.
A group or joint activity.	An individual activity.
The goal is well motivated from the learner's point of view.	The goal is not well motivated from the student's point of view.
The activity is captivating—fun.	"Fun" is not a relevant consideration.
No interruptions are permitted.	There are continual interruptions.
There are frequent "flow" experiences.	There are seldom any "flow" experiences.
The activities are self-paced.	The activities are fixed, forced-paced.
The person has a choice of topic, time, and place.	The topics are fixed, as are the time and place.
The activities can be done throughout life in many environments.	The activities are primarily restricted to ages 6–20+, in a schoolroom.

Classrooms ought to be ideal settings in which to provide the motivation and the information needed for later reflection. In the classroom, one can start the restructuring process, capture attention, and present basic methods.

Neither the classroom nor the fancy teaching system is sufficient. Successful learning requires accretion of knowledge, tuning of the procedures, and considerable reflection in order to restructure one's conceptual understanding. These are best done through a combination of experiential and reflective modes of cognition, optimal cognition, sustained and focused upon the topics at hand, free of distracting interruptions.

Multimedia for education must minimize the fluff and get the users working—working hard, not because they have to but because they want to. Educators must get rid of the lockstep, paced, arbitrary instruction, where students work hard because they have to, not because they want to. Together, we might reach a solution.

THREE

THE POWER OF REPRESENTATION

THE POWER OF THE UNAIDED MIND IS HIGHLY OVERRATED. WITHOUT external aids, memory, thought, and reasoning are all constrained. But human intelligence is highly flexible and adaptive, superb at inventing procedures and objects that overcome its own limits. The real powers come from devising external aids that enhance cognitive abilities. How have we increased memory, thought, and reasoning? By the invention of external aids: It is things that make us smart. Some assistance comes through cooperative social behavior; some arises through exploitation of the information present in the environment; and some comes through the development of tools of thought—cognitive artifacts—that complement abilities and strengthen mental powers.

The limits of the average mind are most easily demonstrated by noting the attention paid to those who have managed to overcome them. We pay homage to those who can remember large quantities of information without any external aid. We pay them money to perform before us on the stage, and we clap delightedly when they learn the names of everyone in the room or tell us the cube root of the serial number of a dollar bill held up by someone in the audience or tell us on what day of the week some arbitrary event a hundred years ago fell. We admire these abilities because they are so unusual and so difficult for the average person to perform. Actually, these skills are difficult even for the expert. They take years to perfect; they require the memorization of numerous tables and word lists, and the learning and continued practice of the computational and mnemonic algorithms. More important, however, is that these are unessential skills. The rest of us live quite productive lives without ever acquiring them. We substitute paper and pencil for mnemonic skills, pocket calculators for computational skills, and printed calendars and tables for extensive memorization and mental calculation.

Probably the most important of our external aids are paper, pencil, and the corresponding skills of reading and writing. But because we tend to notice the unique, not the commonplace, few recognize them for the powerful tools that they are, nor does the average person realize what breakthroughs in reasoning and technology were required to invent writing, numerical representations, portable and reliable pens and pencils, and inexpensive, functional writing paper.

Oral cultures, societies that do not yet have a written language and that also lack the mechanical tools of technological cultures, do not share the benefits. These cultures have not developed advanced mathematics or formal methods of decision making and problem solving. The society that does not yet have writing also has less formal schooling. Instead, most education is conducted through apprenticeships, by watching, copying, and being guided by those who know how to do the task being learned. Their need for formal schooling is limited. They haven't developed mathematics or science, formal history, or extensive commercial records because they can't without the aid of artificially constructed artifacts. It is *things* that make us smart.

Two thousand years ago, Plato wrote the collected dialogues in which he presented the views of Socrates on the important issues of those times. Socrates, Plato tells us, argued that books would destroy thought. How could this be? After all, books, reading, and writing are considered to be the very essence of the educated, intellectual citizen. How could one of the foremost thinkers of civilization deny their importance?

Socrates is famous for his dialogues between teacher and student in which each questions and examines the thoughts of the other. Questioning and examination are the tools of reflection: Hear an idea, ponder it, question it, modify it, explore its limitations. When the idea is presented by a person, the audience can interrupt, ask questions, probe to get at the underlying assumptions. But the author doesn't come along with a book, so how could the book be questioned if it couldn't answer back? This is what bothered Socrates.

Socrates was concerned with reflective thought: the ability to think deeply about things, to question and examine every statement. He thought that reading was experiential, that it would not lead to reflection.*

> SOCRATES: *Then anyone who leaves behind him a written manual, and likewise anyone who takes it over from him, on the supposition that such writing will provide something reliable and permanent, must be exceedingly simple-minded; he must really be ignorant of Ammon's utterance, if he imagines that written words can do anything more than remind one who knows that which the writing is concerned with.*
>
> PHAEDRUS: *Very true.*
>
> SOCRATES: *You know, Phaedrus, that's the strange thing about writing, which makes it truly analogous to painting. The painter's products stand before us as if they were alive, but if you question them, they maintain a most majestic silence. It is the same with written words; they seem to talk to you as if they were intelligent, but if you ask them anything about what they say, from a desire to be instructed, they go on telling you just the same thing forever. And once a thing is put in writing, the composition, whatever it might be, drifts all over the place, getting into the hands not only of those who understand it, but equally of those who have no business with it; it doesn't know how to address the right people, and not to address the wrong. And when it is ill-treated and unfairly abused it always needs its parent to come to its help, being unable to defend or help itself.*
>
> PHAEDRUS: *Once again you are perfectly right.*

Socrates was an intellectual, and to him thinking was reflection or nothing. He didn't go for this experiential stuff. The worst kind of writing for people like Socrates would be novels, storytelling. A story engages the mind in an experiential mode, capturing the reader in the flow of events. All such experiential modes—music, drama, and novels—were considered to be the entertainment of the masses, not worthy of serious respect. Socrates worried that reading would be too passive, an acceptance of the thoughts of the writer without the chance to question them seriously.

In the Middle Ages, just the opposite was true. Reading was generally done aloud, often to an audience. It was an active process, so active that Susan Noakes, in her analysis of medieval reading, points out "that it had been recommended by physicians, since classical times, as a mild form of exercise, like walking."

Moreover, Noakes observes that the characteristics of a good novel today were unheard of in earlier times: "Today, many readers take as the hallmark of the good novel the way it propels them to read it continuously, without putting it down, from beginning to end. Readers of many late medieval books would have been forced, on the other hand, to read discontinuously, stopping to puzzle over the relationship between complement and text." (The term *complement* refers to the dialogue provided through the illustrations and marginal comments—illuminations and glosses—sometimes put in by the author, sometimes by the copyist, sometimes by other readers.)

During the Middle Ages, readers were taught the rules of rhetoric and were implored to employ them with each sentence: *mnemonics*, to memorize and learn the material; *allegory*, to find the multiple levels of meaning hidden beneath the literal text; *typology*, to think in historical parallels. No text was thought to be complete without mental elaboration in the mind of the individual reader or debates within the social group that might be listening to the read-aloud text.

Readers in the latter part of the Middle Ages did with books exactly what Socrates had claimed was impossible: They questioned and debated each idea. True, the author wasn't around, but in many ways that made the job more challenging, more interesting. Read a sentence, question it. Read a page, criticize it. No authors to object. No authors to refute your arguments with the force of their rhetoric. Readers were free to develop their own objections and opinions without interference from meddling authors. Today we may have regressed to match the fears of Socrates: We read too quickly, without questioning or debating the thoughts of the author. But the fault does not lie with the book, the fault lies with the reader.

Cognitive artifacts are tools, cognitive tools. But how they interact with the mind and what results they deliver depend upon how they are used. A book is a cognitive tool only for those who know how to read, but even then, what kind of tool it is depends upon how the reader employs it. A book cannot serve reflective thought unless the reader knows how to reason, to reflect upon the material.

COGNITIVE ARTIFACTS

The cognitive age of humans started when we used sounds, gestures, and symbols to refer to objects, things, and concepts. The sound, gesture, or symbol is not the thing itself; rather, it stands for or refers to the thing: It represents it.

The powers of cognition come from abstraction and representation: the ability to represent perceptions, experiences, and thoughts in some medium other than that in which they have occurred, abstracted away from irrelevant details. This is the essence of intelligence, for if the representation and the processes are just right, then new experiences, insights, and creations can emerge.

The important point is that we can make marks or symbols that represent something else and then do our reasoning by using those marks. People usually do this naturally: This is not some abstract, academic exercise. Suppose Henri had an auto accident and is describing it to his friends. The description might go something like this:

"Here," Henri might say, putting a pencil on the tabletop, "this is my car coming up to a traffic light. The light is green, so I go through the intersection. Suddenly, out of nowhere, this dog comes running across the street." With this statement, Henri places a paper clip on the table in front of the car to represent the dog. "I jam on my brakes, which makes me skid into this other car coming from the other direction. We don't hit hard, but we both sit there stunned."

Henri takes another pencil and lets it represent the second car. He manipulates the pencil representing his car to show it skidding, then turning and hitting the other pencil. Now the tabletop has two pencils touching each other and a paper clip.

"The dog disappears," Henri says, moving the paper clip off the table. "Then the light turns red, but I can't move. Suddenly, this car comes rushing down the side street. It has a green light, but here we are, stuck right in the middle of the intersection. Boom, it hits us, like this"—and Henri uses his finger to show the third car coming from the side and hitting the two pencils, scattering them.

In this scenario, the tabletop, pencils, paper clip, and finger are all used symbolically. They stand for the real objects—the street, the three cars, and the dog. In the listener's head are other symbols to represent the streets and the traffic light. Notice how difficult it would have been to tell this story without the artifacts, without the tabletop, pencils, and paper clip. In fact, you, the reader, may have had some problems following it in this text unless you tried to visualize the scene in your head: What was the path of the dog? Exactly where were the two cars stopped in the intersection? How did the third car travel?

The story on the tabletop helps show some of the simpler properties of artifacts. You can see how they help the mind keep track of complex events. The same representational structure is also a tool for social communication: Several different people can share the tabletop and the story at the same time, perhaps suggesting alternative courses of action. "Look," Marie might say, picking up one of the pencils, "when you saw the dog, you should have gone like this." "Ah, but I couldn't," Henri might respond, "because there was another car there," and he puts yet another pencil on the tabletop. The tabletop becomes a shared workspace with shared representations of the event.

Note what is now happening: People are using the artifacts themselves to reason about alternative courses of action. The representation substitutes for the real event. A problem, of course, is that the representations are abstractions. The pencil may represent the car, but it doesn't have the correct size or mass. It isn't possible to show how fast the real car was going or how much it would skid if the brakes were applied. All this would require more powerful representations. Nonetheless, the representation adds dramatically to the person's power to describe the event. It enables other people to understand better. It makes it easier to analyze alternative actions. It adds power and precision to the memory of the unaided mind.

A good representation captures the essential elements of the event, deliberately leaving out the rest. Pencils don't look anything at all like cars, yet for the purposes of understanding the incident, that difference doesn't matter. A representation is never the same as the thing being represented, else there would be no advantage to using one. The critical trick is to get the abstractions right, to represent the important aspects and not the unimportant. This allows everyone to concentrate upon the essentials without distraction from irrelevancies. Herein lie both the power and the weakness of representations: Get the relevant aspects right, and the representation provides substantive power to enhance people's ability to reason and think; get them wrong, and the representation is misleading, causing people to ignore critical aspects of the event or perhaps form misguided conclusions.

To understand cognitive artifacts, we must begin with an understanding of representation. A representational system has two essential ingredients, shown in Figure 3.1:*

1. *The represented world:* that which is to be represented;
2. *The representing world:* a set of symbols, each standing for something—representing something—in the represented world.

Representations are important because they allow us to work with events and things absent in space and time, or for that matter, events and things that never existed—imaginary objects and

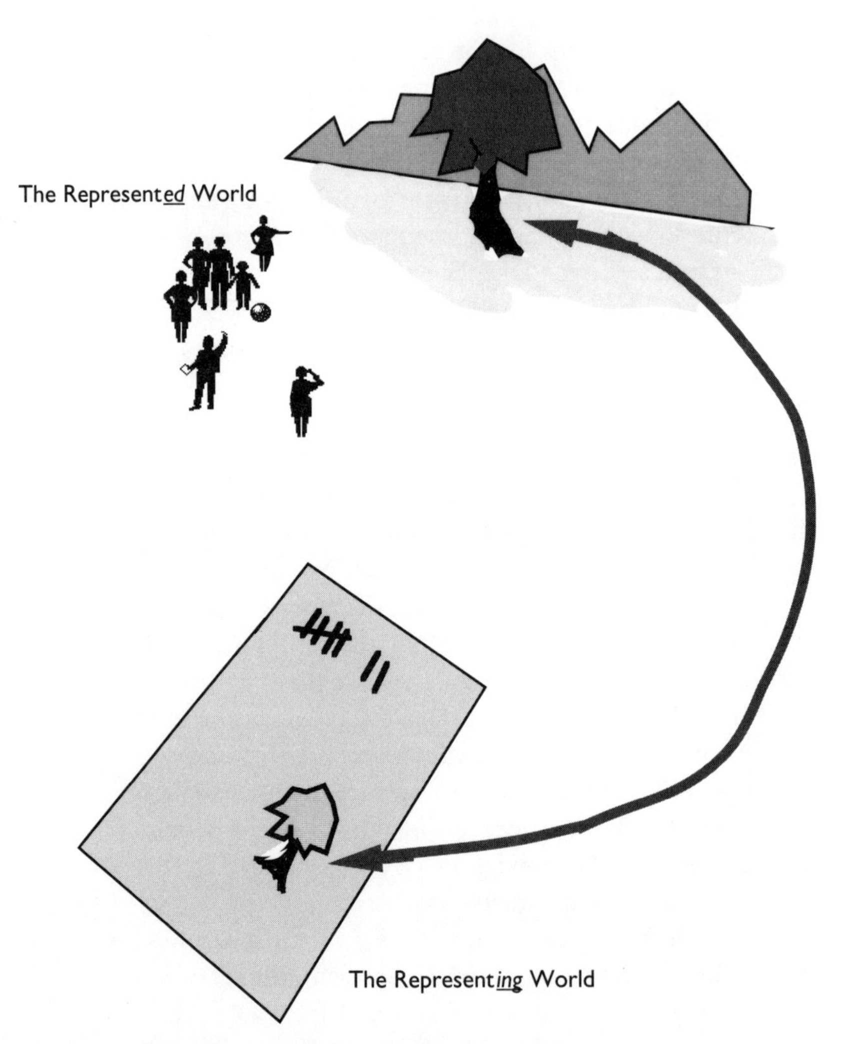

Figure 3.1 The represented and representing worlds. The world to be represented is shown on top—the "represented" world consisting of people, a tree, mountains, and a ball. The "representing" world is shown as marks—symbols—on a sheet of paper. The representing world is an abstraction and a simplification of the represented world. In this example of a representing world, the tally marks each represent one person, and the drawing represents the tree. The other aspects of the real (represented) world are absent from the representing world.

concepts. External representations, especially ones that can be part of a workspace shared with others, require some sort of constructed device to support them: an artifact. Even if the representation is as simple as stones placed in a special arrangement on the ground or a diagram drawn in the sand, its use as a representation is artificial, with a designated space and often with a verbal explanation to interpret for each object in the representing world just what aspect of the represented world it stands for. We have invented more powerful artifacts than sticks, stones, and sand, of course—artifacts that support a variety of representations, that are long-lasting, portable, easily reproduced and communicable over distances, and capable of powerful computational abilities in their own right.

The critical property of the representations supported by cognitive artifacts is that they are themselves artificial objects that can be perceived and studied. Because they are artificial, created by people, they can take on whatever form and structure best serves the task of the moment. Instead of working with the original idea, concept, or event, we perceive and think about representations that are better suited to match our thought processes. Figure 3.1 serves as an example of this ability to represent knowledge. The figure is itself a representation, one that represents the concept of representation. It contains a representation of yet another artifact (labeled "The Represent*ing* World") and the symbols on it, as well as the relationship between that artifact and the world that it represents. Hence, the figure is a metarepresentation: a representation of a representation.

This ability to represent the representations of thoughts and concepts is the essence of reflection and of higher-order thought. It is through metarepresentations that we generate new knowledge, finding consistencies and patterns in the representations that could not readily be noticed in the world. These higher-order representations are very difficult for the unaided mind to discover. In principle, it can be done without artifacts, with just the unaided mind, but in practice, the limited ability to keep track of things in active consciousness severely reduces that possibility.

Once we have ideas represented by representations, the physical world is no longer relevant. Instead, we do our thinking on the representations, sometimes on representations of representations. This is how we discover higher-order relationships, structures, and consistencies in the world or, if you will, in representations of the world. The ability to find these structures is at the heart of reasoning, and critical to serious literature, art, mathematics, and science. The ideal, of course, is to develop representations that

- Capture the important, critical features of the represented world while ignoring the irrelevant
- Are appropriate for the person, enhancing the process of interpretation
- Are appropriate for the task, enhancing the ability to make judgments, to discover relevant regularities and structures

There are many kinds of artifacts. Experiential artifacts have different functions from reflective ones. Experiential artifacts provide ways to experience and act upon the world, whereas reflective artifacts provide ways to modify and act upon representations. Experiential artifacts allow us to experience events as if we were there, even when we are not, and to get information about things that would be inaccessible, even if we were present. A telescope gives us information about something distant in space. A movie or recording lets us experience events distant in time and space. Instruments, such as the gas gauge of an automobile, give us information about states of equipment that would otherwise be inaccessible. Experiential artifacts thus mediate between the mind and the world.

Reflective artifacts allow us to ignore the real world and concentrate only upon artificial, representing worlds. In reflection, one wants to contemplate the experience and go beyond, finding new interpretations or testing alternative courses of action. The process can be both powerful and dangerous. The power comes from the ability to make new discoveries. The danger occurs whenever we fool ourselves into believing that the representation is the reality.

When we concentrate only upon the information represented within our artifacts, anything not present in the representation can conveniently be ignored. In actuality, things left out are mostly things we do not know how to represent, which is not the same as things of little importance. Nonetheless, things not represented fall in importance: They tend to be forgotten or, even if remembered, given little weight. This is the lesson of Chapter 1: We value what we can measure (or represent).

MATCHING THE REPRESENTATION TO THE TASK

> Solving a problem simply means representing it so as to make the solution transparent. *(Simon, 1981)*

Let's play a game: the game of "15." The "pieces" for the game are the nine digits—**1, 2, 3, 4, 5, 6, 7, 8, 9.** Each player takes a digit in turn. Once a digit is taken, it cannot be used by the other player. The first player to get three digits that sum to 15 wins.

Here is a sample game: Player **A** takes **8.** Player **B** takes **2.** Then **A** takes **4,** and **B** takes **3. A** takes **5.**

> *Question 1:* Suppose you are now to step in and play for **B.** What move would you make?

This is a difficult problem for several reasons, all traceable to the way I described the problem—to the representation. The task is described as a problem in arithmetic. To figure out what move to make, you have to consider what possibilities both you and **A** have for winning. This requires a lot of calculation to determine which triples of digits sum to 15. There are few aids to memory, so it is difficult to keep track of which player has chosen which digits, which ones remain. I have deliberately presented the game information to you in a representational form that is awkward to use: The moves are listed sequentially, making it difficult to see just which digits **A** and **B** each have. Although the arithmetic is simple, keeping track of all the possibilities while doing the arithmetic makes the game difficult.

Now let's play a different game, this one the children's game of ticktacktoe (also called "naughts and crosses" and "three in a row"). Players alternately place a naught (the symbol **O**) or a cross (the symbol **X**) in one of nine spaces arranged in a rectangular array (as shown in the following illustration). Once a space has been taken, it cannot be changed by either player. The first player to get three symbols in a straight line wins. Suppose player A is X and B is O, and the game has reached the following state:

X	O	X
	X	
O		

Question 2: Suppose you are now to step in and play an O for B. What move would you make?

Unlike the game of 15, this time the task is easy. This is a spatial game, not one of arithmetic. To see what is happening, just look at the board: A quick glance shows that A is all poised to win (by completing a diagonal line of Xs) unless blocked by an O in the lower right-hand corner.

Question 1 was hard because the game of 15 requires reflection, with few external aids. Question 2 was easy because it could be answered experientially, perceptually: No computation required—just look at the board and see the proper move.

But note, the two games are really the same. If you think of the nine digits of the game of 15 arranged in a rectangular pattern, you see that it is identical to the game of ticktacktoe:

4	3	8
9	5	1
2	7	6

Remember the moves in the game of 15? A had selected 8, 4, and 5; B had selected 2 and 3:

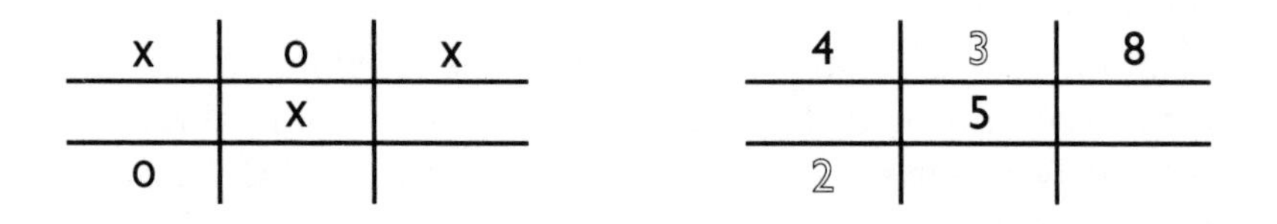

Player B, you, had better select the digit 6, in the lower right corner.

The two games are what we call "problem isomorphs" (from the Greek *iso*, for "the same," and *morph*, for "form").* Technically, questions 1 and 2 are identical, but as the example shows, the choice of representation changes the task and the difficulty dramatically.

Although the spatial representation of ticktacktoe is much easier for people to play than the arithmetic one of 15, for computers the arithmetic representation is much easier. A computer program to solve ticktacktoe spatially would have to figure out whether the Xs and Os were on a straight line: It would have to solve the trigonometric relationships among the points. How much easier for us, since we can simply look and see: The human perceptual system is designed for this task. We find the method used by the computer difficult and cumbersome, although we are quite capable of programming the computer to follow the method. In return, it is very difficult for the computer to do the perceptual processing.

This example illustrates two points. First, the form of representation makes a dramatic difference in the ease of the task, even though, technically, the choice does not change the problem. Second, the proper choice of representation depends upon the knowledge, system, and method being applied to the problem. In this case, the method hardest for the human is easiest for the computer, and the method easiest for the human is hardest for the computer. The example therefore also illustrates the differing yet complementary powers of human and computer information processing.

The power of a representation that fits the task shows up over and over again. Bad representations turn problems into reflective challenges. Good representations can often transform the same problems into easy experiential tasks. The answer so difficult to find using one mode can jump right out in the other.

Consider the task of planning an airline trip between two cities. Suppose I want to travel from my hometown of San Diego (California, U.S.A.) to London (England, U.K.). The way in which airline information is typically presented is shown in the accompanying table: the format employed by the *Official Airline Guide* (the OAG), perhaps the most widely used source of airline information for professional travelers within the United States.

	1131	SAN	0820+1	LGW	AA	2734	FCYBM	D10	1
		AA	2734	CHG PLANE AT DFW					
X12	1805	SAN	1425+1	LGW	BA	284	FJMSB	D10	1
	2100	SAN	2030+1	LHR	TW	702	FCYBQ	*	2
		TW	702	EQUIPMENT 767 LAX-L10					

This excerpt from the *Official Airline Guide Worldwide Edition* (November 1990) shows three flights between San Diego and London. Reading left to right, the top line shows a flight leaving at 11:31 AM from San Diego (SAN) and arriving at 8:20 AM the next day (the +1) at London's Gatwick airport (LGW). This is American Airlines flight 2734, with five classes of service (FCYBM), using a DC-10 and making one stop. The second line states that the flight has a plane change at the Dallas/Fort Worth (DFW) airport. The third line shows a flight that goes every day except Monday and Tuesday (X12): British Air flight 284, with one stop. (The arrival time, 1425, is given in European, twenty-four-hour time: 1425 is 2:25 PM.) The fourth line is a TWA flight that makes two stops and lands at London's Heathrow airport (LHR) at 8:30 PM, and the last line indicates that between San Diego and Los Angeles (LAX), the flight is on a Boeing 767, but from Los Angeles, it will be a Lockheed L-1011.

The OAG's presentation is designed to pack as much information as possible into the smallest amount of space. The monthly worldwide edition is printed in tiny type on over fifteen hundred large pages. Although the publishers have done a creditable job of making the entries usable, the user still has to do considerable mental processing and copying of information. The publishers have unwittingly transformed the selection of a flight into a reflective task.

Suppose my desire is a flight that arrives in London late in the afternoon. At first glance, the OAG format would appear to be perfect because column four shows arrival time directly: I need only scan the arrival times for the one most convenient. This would suggest the TWA flight that leaves San Diego at 9:00 in the evening and arrives in London at 8:30 in the evening the next day. Wonderful: I get on the plane, read a book, have a brief sleep, and when I get to London, clear customs, and get to my hotel, it is time for bed.

But is this true? Closer reading indicates that I had better not go to sleep right away: There is a plane change at Los Angeles. And there are two stops: Los Angeles and where? Is this flight longer than the others?

While I want to arrive late in the afternoon, I do not want to spend several extra hours in traveling. So let me see which flight has the shortest duration. Now we see how the display affects the task: It was easy to search for a flight by arrival time, but it is not so easy to find a flight by duration. I have to do some arithmetic, subtracting departure times from arrival times, which is not easy given that they are on different days and that there is a seven-hour time difference between the two cities.*

To do the arithmetic, I must invent an intermediate notation. Whenever an arrival occurs on the day following departure, add twenty-four hours to the arrival time: 1:00 AM the next day is 25:00; 8:30 PM the next day is 44:30 (20:30 + 24:00). This puts arrival times in a format that makes it easy to find the differences in time between departure and arrival, and when I subtract the seven-hour time difference, I get the following durations:

Flight	**Depart**	**Arrive**	**Difference**	**Duration** (Diff. −7)
AA 2734	11:31	32:20	20:49	13:49
BA 284	18:05	38:25	20:20	13:20
TW 702	21:00	44:30	23:30	16:30

The TWA flight takes almost three hours longer than the others, so even though it arrives at a good time, the tradeoff of an extra three hours travel time is not acceptable. Note that this new arrival notation makes it easier to do the arithmetic but harder to figure out what time the flights arrive. One table makes it easier to choose the shortest flights; the other table makes it easier to check what time the flights arrive. Of course, I could simply add an extra column to the first table giving duration, but in the crowded pages of the OAG, there is simply no room for any information that can be derived.

All of this comparing and planning is reflective. I look at the information given in the OAG and ask questions of it, restructuring the information and performing new computations. This is an excellent example of the power of reflection, except it shouldn't be needed. A different form for presenting this information would change the task to an experiential one, where the answers would appear through inspection.

The OAG uses a table to present its information, and this made some of the comparisons difficult. Stephen Casner has shown how the graphic presentation of scheduling information can simplify some of the decision making in flight planning.* So, borrowing from his work, let us examine these three flights.

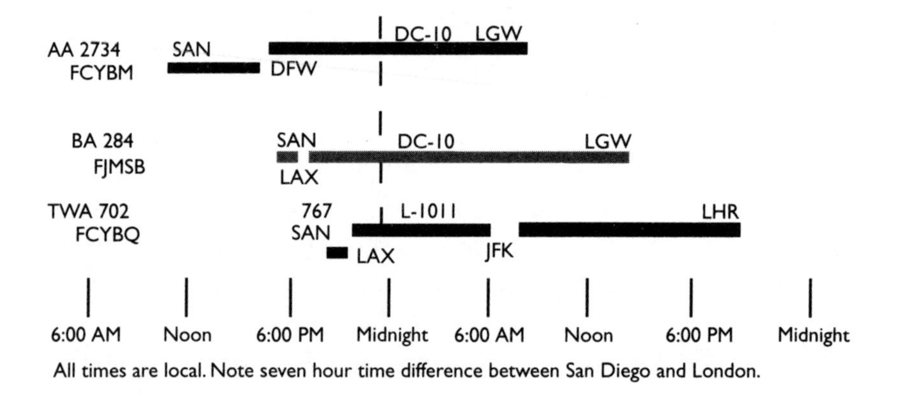

All times are local. Note seven hour time difference between San Diego and London.

This graphic display does appear to make some of the information about the flights much easier to comprehend. It shows all the information in the OAG display, plus more information about the stops. Durations of the three flights are indicated by the lengths of the lines connecting departure and arrival times. The notation also provides a simple way to represent plane changes (the "steps" in the lines) as well as the amount of time spent at stops (the gaps in the lines). The AA flight has a stop, plane change, and delay at Dallas/Fort Worth (DFW). The BA flight stops with a delay but no plane change at Los Angeles (LAX). The TWA flight has a plane change at Los Angeles and a stop with no plane change but a long layover in New York at Kennedy airport (JFK).

Which is the shortest-duration flight? We have already discussed the difficulties of answering this question from the OAG table. In theory, the answer should be easy to discover in the graphic display because all that needs to be done is to compare the lengths of the three lines. In practice, as you can readily see for yourself, the comparisons are not so easy to make. To compare flight durations, you must mentally line up the lines to determine which is the shortest. This example shows that perceptual processing alone does not guarantee success. Whenever mental transformations are necessary in order to make comparisons of the configurations, graphic representation presents the viewer with a difficult task.

The comparison is finally transformed into an experiential task by lining up the starting points: Now you can just look and immediately see the answer.

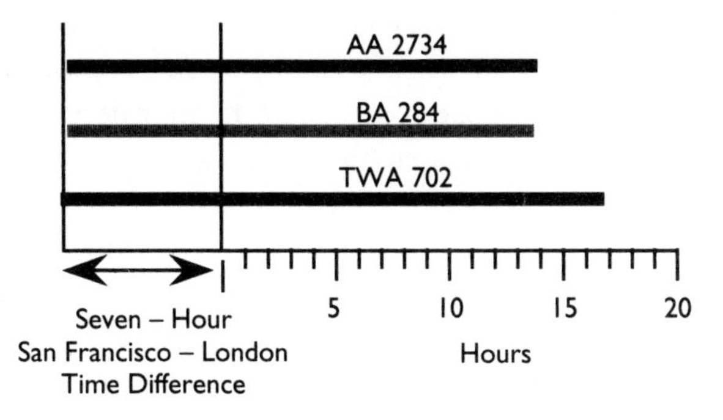

The TWA flight is the longest, and the other two have approximately the same duration. Line up the starting points, remove some distracting clutter, and we have an easy task: The task that used to require arithmetic in the table or mental superposition of lines in the other graphic display can now be done by simply scanning the diagram to find the line that sticks out most (for the longest flight) or least (for the shortest flight).

This new representation also has another advantage. Because the flight times are given in local time, the flight duration is seven hours less than the lengths of the lines would suggest. To determine the actual amount of time on the airplane, you have to subtract the seven-hour time difference. But with this new graph, even the subtraction task is easy. We simply need to move the starting point for the comparison of the lines seven hours to the right, as illustrated on the diagram.

What do we conclude about the appropriate representation for a task? The answer depends upon the task. To know the class of service or the type of airplane, text is superior. To know the exact minute of departure (11:31, say), the printed number is needed. To make a rapid comparison of flight duration, the graphic display is best.

Now that we have seen how graphic displays can simplify the task, what should the OAG do? I recommend that it continue as it is. The publishers of the OAG have a different task from the users. They need to make available as much relevant information as possible. Space is clearly of great importance, and the textual presentation the OAG provides is both efficient and relatively usable. The OAG has changed its format over the years to improve the usability. The graphic display takes up much more space than the tabular one. The most appropriate format depends upon the task, which means that no single format can ever be correct for all purposes.

Someday, not too far in the future, all the information will be available on electronic devices whose displays will allow the same information to be presented in a variety of ways: different layouts for different needs. Wouldn't it be nice to be able to see a listing of all flights organized by time of arrival or by duration of flight or by price of the ticket? Displays that let us switch instantly from numerical format to graphic, depending upon the task? And that let us move among all the formats until all the information needed was available, neither too much nor too little?*

HOW REPRESENTATIONS AID INFORMATION ACCESS AND COMPUTATION

There are two major tasks for the user of an information display:

1. Finding the relevant information;
2. Computing the desired conclusion;

In our examination of information displays, we can note what kinds of assistance the displays provide for these two aspects: What aids are given to help the person's access to the appropriate information? What aids are given to help with the computations?

Consider this example, taken from the draft of a Ph.D. dissertation:

> They found that while subjects would rate the analogies, from best to worst, as literally similar, true analogy, mere appearance, and false analogy, their recall for stories, from best to worst, was literally similar, mere appearance, true analogy, and false analogy.

Why is the sentence so unintelligible? Just consider what you have to do to figure out what it means:

> Best to worst, um, best for analogies is literally similar. And stories, best is literally similar. Gee, those are the same. Let's see, next best for analogies is, um, true analogy. Next best for, um, stories, is, um, mere appearance. Hmm, that's different.

The task of understanding the sentence is an example of reflective thought, unnecessarily reflective, for the information in the sample sentence can also be displayed in a chart like this:

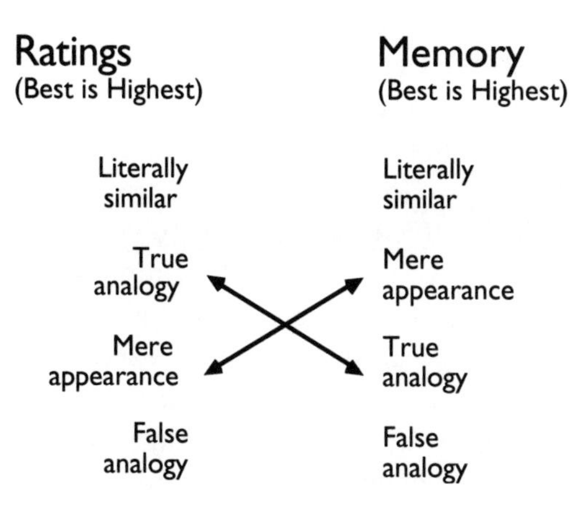

This diagrammatic format uses several techniques to aid the reader:

Needs of the Reader	Provided By
Finding critical comparisons	The lines with arrowheads make the significant comparisons easy to find.
Finding the relevant variables to be compared	Lining up the items.
Remembering the ranking of conditions	Ordering them vertically—the higher, the better.
Comparing the different conditions	Putting the four conditions into two vertical columns, lined up horizontally.
Search and computation	Lining up the right borders of the left list and the left borders of the right list.

The diagram contains exactly the same information as the original, written sentence, but in a form much easier to understand. The tabular arrangement has made both the search and the computations—which, in this case, are comparisons—simpler. Is this a graphic, a chart, or a table? It doesn't matter: It uses an appropriate display format for the task.

Example: Medical Prescriptions

Medical prescriptions are growing ever more complex, with many people being required to take numerous medications daily. How well do people cope with following their prescriptions? Not very well. Several surveys have shown that between 10 percent and 30 percent of the people studied were unable to determine how much medication they should take at any time. In one study, arthritic patients were asked to bring their medication to the experimenters and then, with the bottles and containers in front of them, to write down their daily medication. They were allowed as much time as needed. The results showed great difficulty in doing the task, with an average error rate of about 14 percent. It should hardly be a surprise that the more medications prescribed, the greater the percentage of error. Those people who were prescribed the largest amounts (seven or more drug dosages a day) made both the highest absolute number of errors and the highest percentage of errors: slightly over 30 percent.

The problems of keeping track of medication are well known. In my local drugstore, several different memory aids are available, all aimed at making it easier to keep track of pill taking. All of them are "pill organizers," boxes divided into compartments labeled by time of day, day of week, or both day and time. In principle, these should be beneficial to patients, once the pills are loaded into the proper compartments. Alas, loading the boxes is not very easy. The boxes do not overcome the fundamental problems of interpreting the prescriptions.

The same study that revealed the 30 percent error rate in taking pills also examined how well patients could use these organizers. Again, the answer is not very well. One patient put twice the recommended medication into one of the boxes. Another box tended to be loaded properly, but the average loading time was over nine minutes! These organizers do not appear to work, not when they still lead to errors or when they require so much time to be loaded with pills.

This is an area crying out for help. Solutions, to be effective, must include and support the needs of all the people involved with the prescription: the patient, the physician and physician's aides, and the pharmacist. This issue can truly be a matter of life or death.

One of the problems is that the prescriptions themselves are not written from the patient's point of view. Consider the following medical prescription from the work of psychologist Ruth Day, a prescription that was given to a patient following hospitalization for a mild stroke.

Inderal	—1 tablet 3 times a day
Lanoxin	—1 tablet every a.m.
Carafate	—1 tablet before meals and at bedtime
Zantac	—1 tablet every 12 hours (twice a day)
Quinaglute	—1 tablet 4 times a day
Coumadin	—1 tablet a day

This set of instructions is very difficult to follow. Speaking of the patient, Day reports:

> Over the next few days, he had difficulty remembering what pills to take, as well as what pills he had already taken. It would be easy to blame the patient: after all, he was 81 years old and had just had a stroke. However, he was highly intelligent, was still working full time (and had even begun a new and demanding career a few years earlier), was not otherwise disoriented, and was highly motivated to return to work and an active life style. (Day, 1988, p. 276)

The physician's list, as presented here, is neatly organized, precise, and easy to read. It is very similar to the format used for most prescriptions in the United States. The problem is that it is set up for the wrong task. The representation is appropriate from the point of view of the prescribing physician: Figure out what the patient needs and write it down. But it simply does not lend itself to usage. The list is organized by medicine, which makes it easy for the physician and the pharmacist to look for any medication and see how it was prescribed. But the patient needs it organized by time: Given the time of day, what actions should be performed? Day tested the usability of the prescription by having people try to answer the following two questions:

1. It is lunchtime (noon). Which pills should you take?
2. If you leave home in the afternoon and will not be back until breakfast time the next day, how many pills of each type should you take along?

As you can determine for yourself, it is not easy to answer these questions. The problem is that following this prescription is a reflective task, when it should be an experiential one. Reflection requires mental effort, something a sleepy, ill patient is apt to have trouble with. To fit the needs of the patient, the prescription should be organized by time of day. Note this organization is still appropriate for the physician or pharmacist. Here is Day's suggested presentation of the information:

	Breakfast	Lunch	Dinner	Bedtime
Lanoxin	✔			
Inderal	✔	✔	✔	
Quinaglute	✔	✔	✔	✔
Carafate	✔	✔	✔	✔
Zantac		✔		✔
Coumadin				✔

Notice that with Day's solution, the items can be organized by time of day (the columns) or by medication (the rows). The users simply scan the list by whichever starting point they prefer. A simple change in representation transforms the earlier, difficult reflective task into a much simpler experiential one. Day's experiments showed that the matrix form was not only easier but also conducive to more accurate interpretation than the original (and more common) format.

As Day points out, the matrix has major advantages over lists. Lists are organized by one factor (medication name, in this example). Matrices allow several different dimensions to serve as organizational keys: in this case, medication name or time of day. Whenever several different needs have to be met, a matrix is apt to be superior.

The matrix organization aids both search and computation. In the original prescription, in order to answer the question "How many pills are taken at lunchtime?" the entire list had to be read and then interpreted. The computations were reasonably extensive, even if simple in nature. With the matrix, the computation merely involves scanning down the "Lunch" column and counting. Once again, the proper choice of cognitive artifact aids the task by transforming it from reflection to experiencing, simplifying the operations that must be performed to reach the desired answer.

REPRESENTING NUMBER

Imagine trying to multiply using Roman numerals—say, CCCVI times CCXXXVIII. It's possible, but very difficult. The same numbers written in modern notation—306 times 238—present an easier challenge. The modern Arabic notation lends itself to efficient algorithms for arithmetic, although to multiply these three-digit numbers will require writing something down. In Roman numerals, each symbol stands for a quantity, and in their original form (where *4* was written as "IIII" and *9* as "VIIII"), it doesn't even matter in what order you write the symbols: CCXXXVIII is the same quantity as ICVXIICXX.* With our modern Arabic numbers, we also use the same symbols repeatedly, but the meaning of each symbol depends upon its location. That's why we need the *0* in 306: The *3* means "300" only in the third position from the right. Roman numerals had no need for a zero.

The choice of representation for numbers makes a big difference in how easy or hard it is to do certain operations. Arabic numbers are not always the best choice for representation.

One of the oldest forms of representing numerical quantities—tally marks—is still the best form when we need a way of counting something rapidly. To count an item, I make a short vertical line, I; adding a second one, II; a third, III; and a fourth, IIII: one new mark for each new item.

Tally marks are easy to make and easy to compare, which is why they are still in use today. Roman or Arabic numerals are much more difficult. Why? Because tally marks are *additive:* With an additive representation, if I wish to increase the value of a previous symbol, I simply add extra marks to the symbol already there. Thus the symbol for 3 (III) readily becomes the symbol for 4 (IIII). Nothing already present has to be changed.

Contrast this with Arabic numerals, which are *substitutive*: With a substitutive representation, if I wish to increase the value of a previous symbol, I must substitute a new symbol for the previous one. To increase the value by 1, I have to cross out the previous value and write the new one. The symbol 1 becomes ~~1~~ 2, and then ~~1~~ 2 becomes ~~1~~ ~~2~~ 3.

Of course, there are other differences between Arabic and tally representations besides the ease of making the marks. Arabic numbers are harder to make than tally marks, but easier to read and to use for computations. To make it easier to read tally marks, we usually modify them somewhat, so we group them into fives, generally like this: IIIII.

Additive notations have another important property: The size of the representation is proportional to the value of the number. So tally marks also serve as a graph. (See Figure 3.2)

These examples show that changes in representation often provide us with tradeoffs: One aspect of the task gets easier while another gets harder. Thus, while counting, tally marks are easier to make than Arabic numbers and easier to compare, especially if the number of objects is relatively small. But for doing calculations, tally marks are much harder to use than Arabic numbers.

Addition Is Easier with Roman than Arabic Numerals

Strange as it may seem, it is easier to add two numbers using Roman numerals than using our everyday Arabic numerals. Today students have to learn the arithmetic table: They start by learning the ten arbitrary symbols for the ten digits, then learn place notation to know that 46 is the same as four 10s plus six 1s. Next, they must memorize the sums for the forty-five possible pairs of numbers. (The ten digits, 0 through 9, have 100 possible combinations. But because of the property called reflexivity—e.g., 4+5 = 5+4—and the ease of adding zero, only forty-five combinations need to be learned.) Finally, students have to learn what to do if there is a carry from one column to the other. All this takes a surprisingly long time to learn.

Roman students simply had to learn the Roman characters for digits—seven different characters go from 1 to 1,000 (I, V, X, L, C, D, M). After that, to add two numbers, they simply combined the symbols together and reordered them, all similar symbols together, the symbols with the greatest value on the left. Then they applied some simplification rules (one rule for

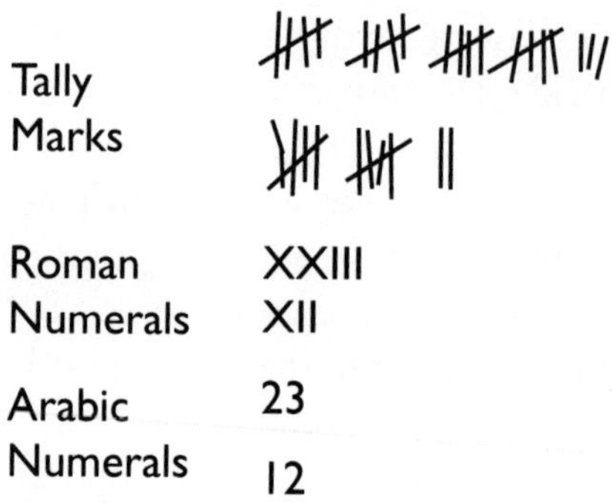

Tally Marks	
Roman Numerals	XXIII XII
Arabic Numerals	23 12

Figure 3.2 Comparing *23* and *12* with tally marks, Roman numerals, and Arabic numerals. Tally marks are an additive representational system in which the length of the representation is proportional to the value being represented. The values of additive representations can be compared experientially. A glance at the figure shows that one value is roughly twice as much as the other. Roman numerals are a modification of tally marks, and so they too can have an additive character, with their length related to the value. A glance at the figure shows that the top value is greater than the bottom one, but the ratios of the lengths of the numerals do not accurately reflect the ratios of the numerical values. Within each place position, Arabic numbers are a substitutive representation, and as this example shows, for small numerical differences, the length of the representation does not provide any information about its value.* The values of substitutive representations have to be compared reflectively, through mental computation.

each symbol) that tell how small symbols combine to make bigger ones (e.g., IIIII = V, VV = X). This is a lot less to learn than the ten symbols of Arabic numerals, the forty-five arithmetic combinations, and the rules for place notation and carry. It's a lot easier too.

Example: 306 + 238

The problem: CCCVI + CCXXXVIII
Combining the symbols: CCCVICCXXXVIII
Reordering the symbols: CCCCCXXXVVIIII
Simplifying gives the answer: DXXXXIIII

The answer, in Arabic numerals: 544

No arithmetic sums have to be known, just how to combine, reorder, simplify, and read the symbols. Roman children had it easier than today's—at least, until they tried to multiply or divide.

Additive and Substitutive Representations

The distinction between additive and substitutive dimensions is important, one that makes a big difference in the ease of understanding graphic representations. The distinction is not well respected by many graphic designers.

Look at Figure 3.3, my redrawing of a chart that was published in a newspaper. The chart uses different kinds of shading superimposed on a map of the United States to indicate what percentages of homes exceed the recommended level of radon, a radioactive gas that we all wish to avoid. Alas, the chart uses the wrong representation: A substitutive representation (different types of shading) is used to represent additive information (percentage of homes that exceed the recommended level of radon). Look at that graph and try to figure out where in the United States radon is most prevalent, least prevalent, and at an average value. The task is hard because the shadings are arbitrary: You have to keep going back to the legend to remember whether a particular shading represents a greater or lesser value than another. The choice of shading transforms this into a reflective task when it should be experiential.

The proper way to draw the figure is to use an ordered sequence of density (an additive scale) to represent percentages (an additive dimension). Try the same task (to determine where radon is most prevalent, least prevalent, and at an average value) with the map shown in Figure 3.4.

I have deliberately introduced a problem with the representation in Figure 3.4 to emphasize the point about the importance of representational format. If you look at the map, it appears that the northwest part of the United States has a very low concentration of radon. That's because that portion of the map is white,

Established % Homes that Exceed
EPA's Recommended Level for Radon

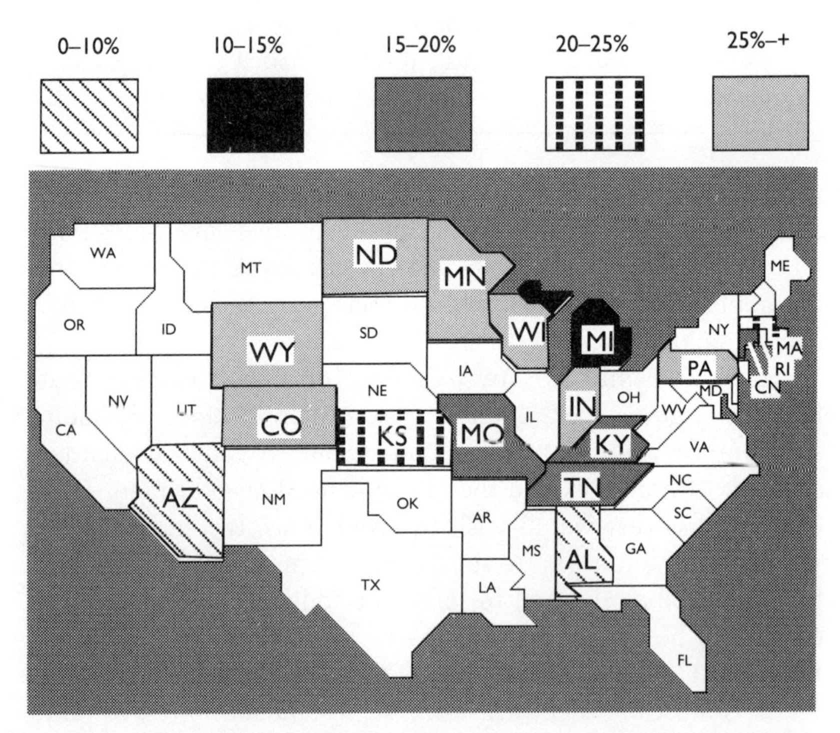

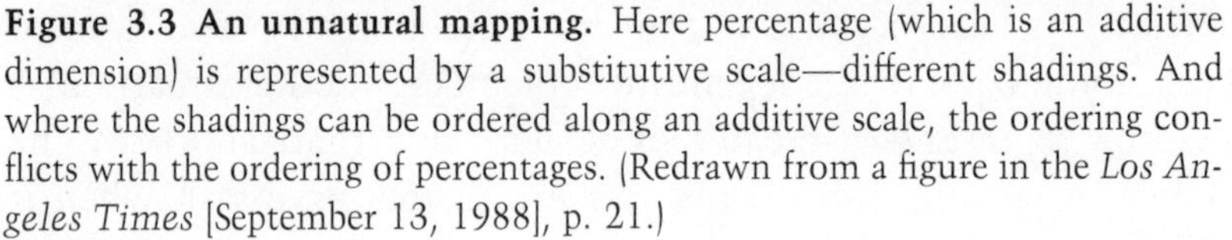

Figure 3.3 An unnatural mapping. Here percentage (which is an additive dimension) is represented by a substitutive scale—different shadings. And where the shadings can be ordered along an additive scale, the ordering conflicts with the ordering of percentages. (Redrawn from a figure in the *Los Angeles Times* [September 13, 1988], p. 21.)

and on the scale of density, white falls to the left of (less than) the 0–10% density. In this case, however, white actually represents those states for which there are no data. A better way to make this graph would be to delete the names of states for which there is no information. I left them in because the natural misinterpretation helps make the point about the impact of representational format.

Figure 3.3, with an inappropriate use of substitutive shading to represent additive percentages, makes the comparison task one of reflection. Figure 3.4, which uses an additive representation of

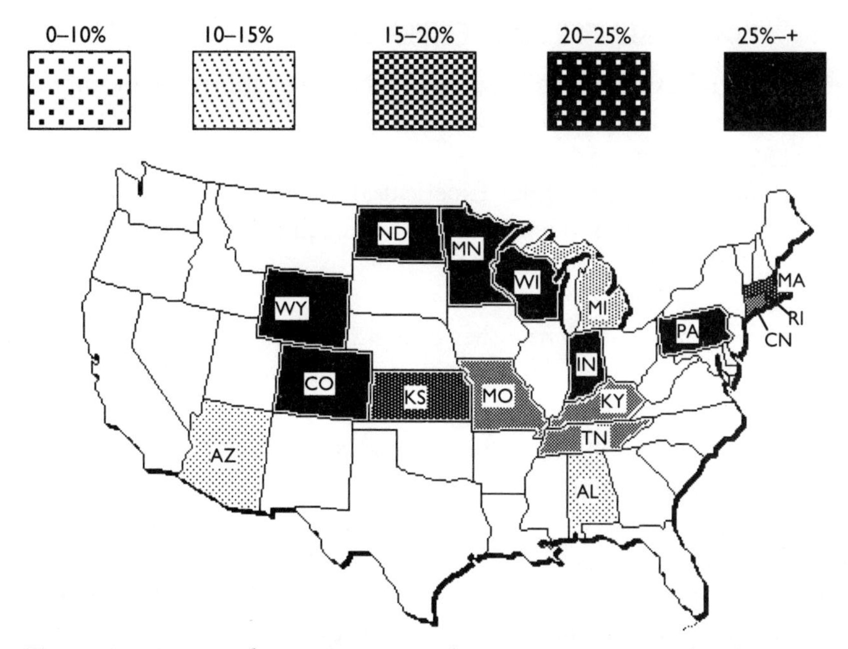

Figure 3.4 A natural mapping. Here the map in Figure 3.3 has been redrawn so that percentage (which is an additive dimension) is represented by an additive scale—ordered densities of shading. Now the density ordering matches the percentage ordering. (Redrawn from a figure in the *Los Angeles Times* [September 13, 1988], p. 21.)

shading to represent the additive percentages, allows the task to be performed experientially.

Color (hue) is frequently used to represent density or quantity, especially in geographic maps, satellite photographs, and medical imagery. But hue is a substitutive representation, and the values of interest are usually additive scales. Hence hue is inappropriate for this purpose. The use of hue should lead to interpretive difficulties. Many colorful scientific graphics, usually generated by a computer, use different hues to represent numerical value. These graphics force the viewer to keep referring to the legend that gives the mapping between the additive scale of interest and the hues. Density, saturation, or brightness would provide a superior representation.

NATURALNESS AND EXPERIENTIAL COGNITION

The several examples throughout this chapter illustrate an important design principle—naturalness:*

> **Naturalness principle:** Experiential cognition is aided when the properties of the representation match the properties of the thing being represented.

I return to these and other design principles in Chapter 4, but let us explore some of the implications of the principle. We humans are spatial animals, very dependent upon perceptual information. Representations that make use of spatial and perceptual relationships allow us to make efficient use of our perceptual systems, to think experientially. Representations that use arbitrary symbols require mental transformations, mental comparisons, and other mental processes. These cause us to think reflectively, and although in many cases this is appropriate and necessary, it is more difficult than experiential cognition. It is also subject to error, especially when people are under high stress.

Mappings are the relationship between the format of the representation and the actual things being represented. They are easier, more reliable, and more natural with well-designed perceptual or spatial representations than with abstract representations. This leads to the second principle:

> **Perceptual principle:** Perceptual and spatial representations are more natural and therefore to be preferred over nonperceptual, nonspatial representations, but only if the mapping between the representation and what it stands for is natural—analogous to the real perceptual and spatial environment.

Graphs are often superior to tables of numbers because in a graph, the height of the line is proportional to the value, so you can compare the different values perceptually. If all you have to work with is numbers, then you have to do some mental arithmetic to see the relationships. Graphs are not always superior to tables, mind you: only when the task is appropriate for perceptual judgments.

We have already seen that to decide whether one number is larger than another, tally marks are superior to Arabic notation because the length of the line of tally marks is directly proportional to the value represented. You might very well wonder what the fuss is about here. The comparison of *23* and *12* seems natural and straightforward: *23* is larger than *12*, what's the big deal? The big deal is that numbers are really not natural. They are reflective tools, not experiential ones. Powerful, essential tools for thought, but nonetheless, reflective. When Arabic numerals were first invented, it was only the most highly educated people who could master them, and their use was debated and, in some cases, prohibited. Even today, it takes years of study in childhood to become proficient at arithmetic, years of practice that later on allow adults to regard the comparison as simple and natural. Anything that requires that much study is not natural.

Try these two numerical comparisons:

A: Which number is larger?
284 912

B: Which number is larger?
284 312

Much to many people's surprise, experimental psychologists discovered that people can answer problem A faster than they can B. The time differences are small, small enough that you can't notice it yourself, but large enough to be easily measured through the appropriate experiments. Even though we experience both comparisons as immediate and effortless, B takes more time and effort than A. Why is this? So far, the only answer that accounts for all the findings is that the Arabic numbers are translated into a perceptual image—an additive representation—before the comparison is performed. The greater the perceptual difference, the easier the task.

From a logical point of view, the two problems A and B seem

equally easy. The point to learn from this is that real psychology is not the same as folk psychology or logic. People have their own commonsense views of how their minds work—a folk psychology. Alas, people are only aware of their conscious experiences, which is a mere fraction of what really goes on. Commonsense views of psychological behavior are reasonable, sensible, and in agreement with everyday experience. Logical views are also reasonable and sensible. Both common sense and logic are often wrong.

Here is the perceptual analog of the numerical comparison. In each problem, the lines are drawn to scale so that they match the earlier questions A and B. Try these two graphic comparisons: In A and B, which line is longer?

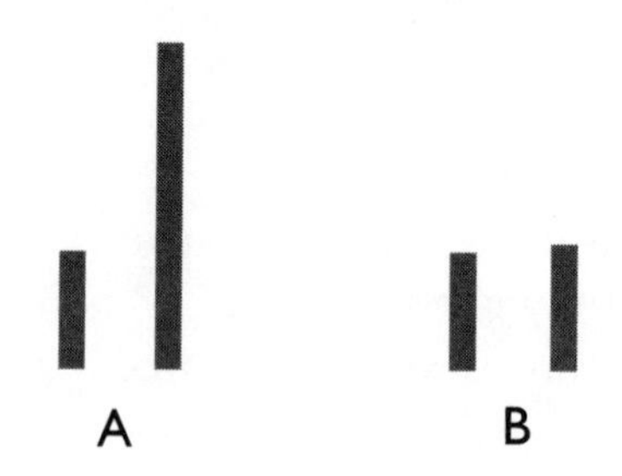

The perceptual comparisons are simple and direct, but here, just as with the earlier questions A and B, comparison A can be done more rapidly than comparison B. But the graphic form of the comparison is easier and faster than the numerical one: The first is experiential, the other reflective. To compare the lengths of two lines, you don't even have to know anything about numbers: The perceptual system handles the chore, simply and efficiently.

Representations that match our perceptual capabilities are simpler and easier to use than those that require reflection. Moreover, under a heavy work load (perhaps under severe stress, danger, and time pressure), representations that require reflection—such as the use of Arabic numbers—are not used as rapidly and efficiently as those that can be used experientially, through simple perceptual comparisons. Where simple comparisons are required, graphic notation is superior. But where exact numerical values are required or where numerical operations must be performed, Arabic notation is clearly superior—that is why it is the standard notation used today.

The power of cognitive artifacts derives from the power of representation. The form of representation most appropriate for an artifact depends upon the task to be performed. The same information may need to be represented differently for different tasks. With the appropriate choice of representation, hard tasks become easy.

FOUR

FITTING THE ARTIFACT TO THE PERSON

YEARS AGO, WHEN I WAS STUDYING HOW PEOPLE REMEMBER THINGS, I used to ask college students to sit in small, soundproof rooms and listen to long lists of spoken digits. After each list, I would "probe" their memory for the digits by presenting a single digit and asking them to tell me what digit had followed the probe in the list. I got lots of interesting data about the limitations of what is now called "working memory" and what, at the time, my collaborator Nancy Waugh and I called "primary memory." I remember how upset I was when I spied one of my experimental subjects writing down the list of digits, then answering the probe question by reviewing the written list. I was so upset that I immediately ordered her out of the room and out of my experiment: She was cheating! She was upset that I was upset: She was getting them all correct—I should have been pleased, she said, not upset.

Today I tend to agree with the student. After all, I had asked her to do a meaningless task, so she had adopted the sensible, intelligent response. Who but an experimental psychologist would expect anyone to remember anything as silly as unrelated digits without the aid of paper and pencil? Better yet, why would anyone ever have to remember such sorts of things without writing them down? The mind is well equipped to retain large amounts of meaningful material, as long as the material has pattern and structure. It is the meaningless, arbitrary stuff of modern life that gives so much trouble. Sure, it is often easier to remember something than to carry around and consult written records for everyday tasks, but why must these things be so meaningless, so arbitrary? Most are arbitrary requirements of today's technology. Most, perhaps all, could be dispensed with through appropriate design. It is quite possible to devise a world in which people learn things because they want to, for convenience and privacy, not because they must. Sensible, meaningful things.

Meanwhile, in our technological, machine-centered world, it really does make sense to remember things by writing them down. Why not? Human working memory is limited, so we can extend it by use of the cognitive artifact. But note: Writing something down doesn't really change our memory; rather, it changes the task from one of remembering to one of writing then, later, reading back the information. In general, artifacts don't change our cognitive abilities; they change the tasks we do.

There are two views of a cognitive artifact: the *personal* point of view (the impact the artifact has for the individual person) and the *system* point of view (how the artifact + person, as a system, is different from the cognitive abilities of the person alone). From a person's *personal* point of view, artifacts don't make us smarter or make us have better memories; they change the task. From the *system* point of view, the person + artifact is more powerful than either alone. Performance of the *system* of person + artifact is indeed enhanced, but that of the individual person is not.

The *personal* point of view:
Artifacts change the task.

The *system* point of view:
The person + artifact is smarter than either alone.

An artifact is not a simple aid. That is, you can't just go out and find some cognitive artifact, and there you are, better at something. Nope, most cognitive artifacts present you with yet another thing to be learned, another manual to be read, another course to be taken, or another period of slow, tedious learning to endure.

Reading, writing, and arithmetic are perhaps our most powerful cognitive skills, but these mental artifacts take years to be learned. Not everyone fully masters them. The study of artifacts is also the study of human capabilities. Why is it so hard for some people to learn these skills? Are there better ways of teaching them? And most important to me, what is it that makes some artifacts effective, others not? Could we develop a science of artifact design that would tell us how to make better artifacts, perhaps ones that were easier to learn and use?

SURFACE AND INTERNAL REPRESENTATION

Once upon a time, before all this electronic and computer stuff came along with its invisible internal representations, we used to be able to see just how our artifacts worked. Everything was physically visible: gears, chains, levers, dials. We could simply move the parts of interest and tell what was going on from their position and motion.

In the modern world of electronic systems, the controls and indicators have almost no physical or spatial relationships to the device itself. As a result, we now have arbitrary or abstract relationships between the controls, the indicators, and the state of the system. This is one reason why these devices are so difficult to learn: Each one uses its own arbitrary choice of operations and methods. The abstraction possible with today's electronic devices means that there doesn't have to be any natural relationship between the appearance of an object and its state.

When a physical file folder is open, it is visibly different from when it is closed. When it is stuffed with paper, it looks different than when it is empty, even when closed. Not so with electronic files. All we can see is whatever the designer thought of providing, which is sometimes a lot, sometimes nothing. The difference is that with the physical folder, the visible properties are an automatic, intrinsic part of its existence, whereas with the electronic folder, any perceivable existence is dependent upon the goodwill and cleverness of its human designer, who provides a perceivable interpretation of the underlying invisible information structures. We understand our artifacts by what is perceivable. With some artifacts, not enough can be perceived.

The natural visibility of artifacts divides them into two broad categories, *surface* and *internal* artifacts. The distinction has important design implications. With surface artifacts, what we see is all there is: They only have surface representations. Take this book. The only information contained here is that represented by the printed words and images: marks on the white paper. The marks are static and passive: They cannot change, unless you physically erase them. With the book and all surface artifacts, what is perceivable is all that exists.

In contrast to surface artifacts there are internal artifacts, in which part of the information is represented internally within the artifact, invisible to the user. Consider, for example, a calculator. What you see is the surface representations, the information visible on the display and the buttons that allow information and instructions to be entered. Beneath those surface representations, however, lie internal representations for the digits and operations, which are unseen by the user but can be manipulated, transformed, and otherwise modified as needed by the calculations being performed. There are even hidden representations—temporary results of the calculations, internal states used only by the calculator that are not displayed, not visible. With the calculator as with all internal artifacts, there is more than can be perceived.

Memory aids such as paper, books, and chalkboards allow for the display and relatively permanent maintenance of representations. The slide rule and abacus are examples of computational devices that only contain surface representations of their information. These devices are primarily systems for making possible the display and maintenance of symbols. I call these "surface representations" because the symbols are maintained at the visible "surface" of the device—pencil or ink marks on paper; chalk on a board; indentations in sand, clay, or wood; and so on. Some of these representations are passive: Once the information has been added, it cannot be changed by the artifact itself. Thus writing, whether on a chalkboard or printed in a book, can be changed by the user, but not by the artifact. Printed tables of reference information have this property. They are meant to be consulted, not to be changed.

Internal artifacts need interfaces, some means of transforming the information hidden within their internal representations into surface forms that can be used. This poses some important design considerations: On the one hand, it offers unlimited possibilities, for the designer can choose whatever representation makes the operation of the artifact best conform to the needs of the user, unconstrained by the physical limitations of the surface representations. On the other hand, it imposes special requirements on the designer, who must now be an expert in both the technology of the artifact and in human psychology, and for artifacts that are used by groups of people, an expert in social interaction as well. Designers never had to think about these issues before: There are few people who can deal with the broad implications of this challenge.

Artifacts that have only surface representations do not need a special interface: The surface representation itself serves as the interface. But this still does not eliminate the need for careful design. There are always alternative designs that make the artifact more or less successful in the fit between its surface representation and the needs of the person and task. Most surface artifacts do have some hidden parts, and the designer must choose which parts to hide, which to make readily available. Still, with a surface artifact, the very nature of the device guarantees some understanding by its users.

Properties of Surface Representations

Some artifacts are passive, incapable of changing their representations without activity by their users. Thus chalkboards and pieces of paper are passive artifacts: Their users initiate all actions that change the surface representations. Some artifacts are active, capable of changing their own representations. Clocks, calculators, and computers are active artifacts, capable of changing their representations without any action by users. A mechanical clock is an active surface artifact; a computer is an active internal artifact.

A person is more like an internal artifact than a surface one: What you see is not all there is. When we interact with one another, we have to transform our thoughts into surface representations so that others can have access to them. This means transforming those ideas into words, facial expressions, gestures, mime, action, sketches, or sounds—exploiting all the sensory capabilities in an attempt to convey our intentions to others. The human surface representations are temporary: Sound fades away, gestures and actions disappear once completed. External surface representations can overcome the limitations of human surface representations. External representations, such as marks and images, can be permanent. This isn't always true, of course: Sounds and video images are transient, lasting only for the duration of the event they signify, but with the proper artifacts, they have the virtue of being forever repeatable.

People and artifacts may have dramatically different internal representations and processing, but the surface representations must be similar or, at least, complementary. A major design problem is to get the surface representations right. Although the surface representations of the artifact must match those of the human, the internal representations need not match and are often most valuable when they differ significantly. But when they differ, the nature of the surface representation is especially critical, for this is where the person gets all the information about the usage and state of the device. Here is where the science of design begins: How shall that information be represented to be of most use?

Some of the properties of simple surface representations are easily seen by asking why a lecturer uses slides or tape recordings. If the lecture is telling of travels to exotic locations, then slides and recordings are necessary because the words of the lecturer cannot convey a complete image of the place being described: The slides and recordings provide better representations—passive, surface, experiential.

What about a lecturer in business or science, where the slides often do not contain any information different from what is also being said? Here slides can serve the cognitive processes of communication in several ways:

- *A shared workspace:* The entire audience can view and reflect upon the same information at once.

- *Cooperative work:* Because there is a shared workspace, everyone can analyze and consider the same points at the same time. Any member of the audience can raise a new question or propose a new insight.
- *Memory permanence:* The slide acts as an external memory, maintaining an accurate record of the words and concepts for as long as the image is projected: The time is under the control of the lecturer as opposed to the vagaries of human memory.
- *Memory quantity:* The slide allows the effective presentation of more information than can be kept active within a person's memory, allowing the viewer to examine different areas selectively, confident that material passed over can be quickly and easily retrieved simply by moving the eye fixation to the appropriate location.
- *Perceptual processing:* The spatial arrangement of ideas helps point out their relationship. The physical presence of the slide helps focus the listener's attention.
- *Individual differences:* Some people prefer auditory information, some visual. Some types of processing are superior when information is auditory, some when it is visual or spatial. The slide provides a redundant communication channel, allowing the listener to select whichever manner of information is easier or preferred.

The appropriate use of a slide assists the audience, for the important concepts remain available for the duration of the slide, not just the duration of the spoken word: The slide acts as an external memory. The speaker can rely on this aspect of the artifact and can refer to critical aspects by pointing to the relevant part of the slide. Even after the slide has been turned off, speakers occasionally point to the place where the information used to be, with the full expectation that the audience will understand what is being referred to, even though nothing is visible on the screen or board.

Slides can also be misused, so much so that some speakers refuse to use them. They can lead to a wandering of attention as the audience reads the slide instead of listening to the speaker. Poor speakers may fail to synchronize their slide presentations with their speech, and if the slides are presented too quickly, the listeners must sometimes engage in rapid copying of the slide material to their own notes, thus forcing them to miss both what is being said and also the contents and implications of the very points they are trying to copy. Finally, far too many government and industrial speakers delight in giving talks with their entire text on the slides, so they read the slides aloud point by point, despite the fact that their audience can read and follow them by themselves faster than the speakers can talk. The result can be a bored and sleepy audience.

The Tower of Hanoi, Oranges, and Coffee Cups

One of my former graduate students, Jiajie Zhang, developed a nice demonstration for his Ph.D. dissertation of how the physical characteristics of an artifact can dramatically change the ease of solving a problem. He did this with one of the favorite puzzles of cognitive scientists, the Tower of Hanoi puzzle, but the result applies to many other domains.

The Tower of Hanoi puzzle is an old familiar one to many people. Figure 4.1 shows the version Zhang studied, slightly modified from the usual puzzle.

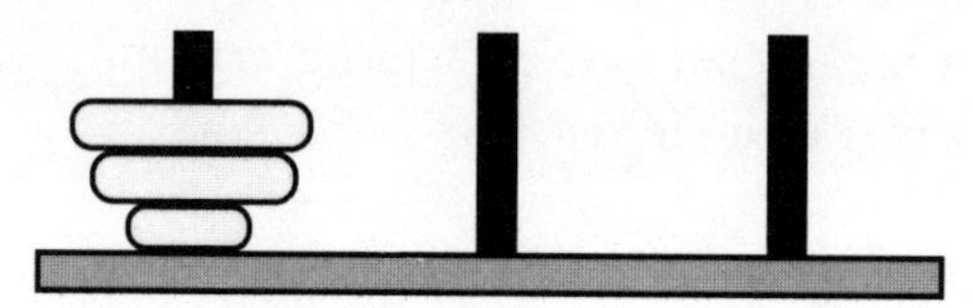

Figure 4.1 The modified Tower of Hanoi puzzle studied by Jiajie Zhang. Three rings are placed on a peg. The goal is to move all three rings from the leftmost to the rightmost peg. Only one ring may be moved at a time, and a smaller ring may not be placed on top of a larger ring. (This is the opposite of the usual convention—usually, the big ring is on the bottom, the small one on top—but this difference doesn't change anything of importance.)

The puzzle consists of three pegs and three rings. At the start, the rings are placed on one peg in order of size (let it be the far-left peg in the drawing). Usually, the rings are arranged so that the smallest is on top, the largest on the bottom. Zhang used the reverse ordering—largest on top, smallest on the bottom—to make this version of the puzzle analogous to "the Coffee Cups puzzle" (to be described shortly). The task is to move the rings from their starting peg to another (the far-right peg in the drawing) following two rules:

Rule 1: Only one ring can be transferred at a time.

Rule 2: A ring can only be transferred to a peg on which it will be the largest.

Zhang studied a variant of the normal puzzle. First, as already noted, the rings were stacked in the reverse order of normal. Second, the ending state was changed to that of one ring per peg in the order (from left to right) large, medium, small, like this:

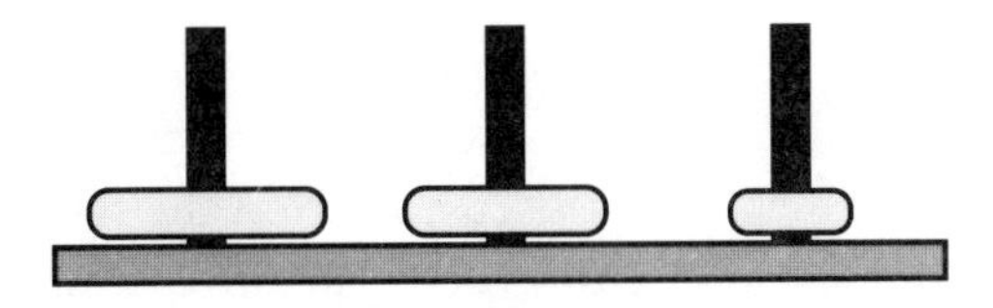

Finally, Zhang added a third rule:

Rule 3: Only the largest ring on a peg can be transferred to another peg.

The third rule makes no difference for the puzzle: If you follow rule 1, then you will only move a single ring at a time, which, given the physical structure of the pegs, means you can only move the top ring. If rule 2 has been followed, the top ring will always be the largest one on a peg. Then why add the rule? Ah, it is unnecessary in this problem only because of the physical structure of the pegs. What if there were no pegs? What if the physical structure did not provide an advantage? This is what Zhang studied: how the physical properties of the puzzle can aid in the solution.

To get at this point, Zhang invented two other versions of the Tower of Hanoi puzzle. All versions had three objects that had to be moved to three different locations; all used the same three rules; and all were formally equivalent—but they varied considerably in their difficulty. The other puzzles are called "isomorphs" of the Tower of Hanoi. An isomorph, you might remember from Chapter 3, is an equivalent problem (*iso* means "equal"), but described differently.

Here are the other two puzzle isomorphs: the Oranges and the Coffee Cups puzzles. (In the descriptions that follow, I use schematic figures to illustrate the puzzles. In his studies, Zhang used real doughnuts for the rings, real plates, and real cups filled with coffee. However, he used three different sizes of balls instead of oranges.)

The Oranges Puzzle

A strange, exotic restaurant requires everything to be done in a special manner. Here is an example. Three customers sitting at the counter each ordered an orange. The customer on the left ordered a large orange. The customer in the middle ordered a medium-size orange. And the customer on the right ordered a small orange. The waitress brought all three oranges on one plate. She placed an empty plate in front of two of the customers and the plate with the three oranges in front of the middle customer (as shown in picture 1).

Because of the exotic style of this restaurant, the waitress had to move the oranges to the proper customers following a strange ritual. No orange was allowed to touch the surface of the table. The waitress had to use only one hand to rearrange these three oranges so that each orange would be placed on the correct plate (as shown in picture 2), following these rules:

Rule 1: Only one orange can be transferred at a time.

Rule 2: An orange can only be transferred to a plate on which it will be the largest.

Rule 3: Only the largest orange on a plate can be transferred to another plate.

How would the waitress do this? That is, you solve the puzzle and show how the waitress has to move the oranges to go from the arrangement shown in picture 1 to the arrangement shown in picture 2.

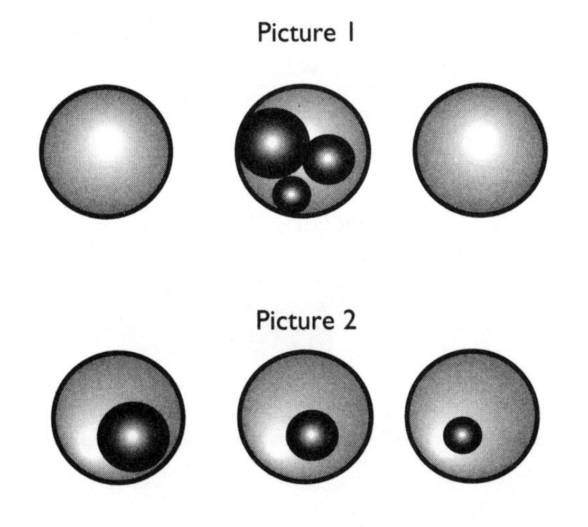

The Coffee Cups Puzzle

A strange, exotic restaurant requires everything to be done in a special manner. Here is an example. Three customers sitting at the counter each ordered a cup of coffee. The customer on the left ordered a large cup of coffee. The customer in the middle ordered a medium-size cup of coffee. And the customer on the right ordered a small cup of coffee. The waiter brought all three cups of coffee on one plate, piled one on top of another. He placed an empty plate in front of two of the customers and the plate with the three cups of coffee in front of the middle customer (as shown in picture 3).

Because of the exotic style of this restaurant, the waiter had to move the cups of coffee to the proper customers following a strange ritual. No cup of coffee was allowed to touch the surface of the table. The waiter had to use only one hand to rearrange these three cups of coffee so that each cup of coffee would be placed on the correct plate (as shown in picture 4), following these rules:

Rule 1: Only one cup of coffee can be transferred at a time.

Rule 2: A cup of coffee can only be transferred to a plate on which it will be the largest.

Rule 3: Only the largest cup of coffee on a plate can be transferred to another plate.

How would the waiter do this? That is, you solve the puzzle and show how the waiter has to move the cups of coffee to go from the arrangement shown in picture 3 to the arrangement shown in picture 4.

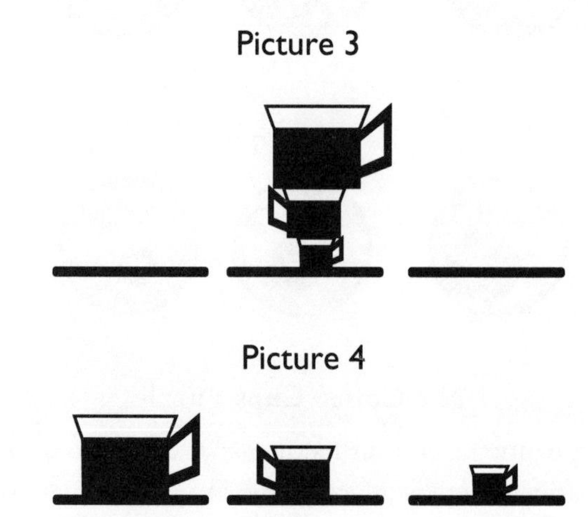

One problem involves a weird way to move oranges about, the other a weird way to move coffee cups. Otherwise, they are the same. Why is this interesting? Because these problems differed greatly in difficulty: The oranges puzzle took almost $2\frac{1}{2}$ times as long as the coffee cups puzzle, with almost twice as many moves and with six times as many errors. The difference had to do with the physical constraints.

In the coffee cups puzzle, although three rules were stated, only one was necessary: rule 1, "*Only one cup of coffee can be transferred at a time.*" Rules 2 and 3 were unnecessary because the cups imposed these rules by their very construction. The cups were filled with real coffee and so constructed that smaller cups would fit inside of larger ones. Rule 3 was not needed because the size of the plate was such that only one cup could fit on it: When more than one cup was on a plate, it had to be stacked on top of the other cups on the plate. Violate rule 2, and because the only way to set more than one cup on a plate was to stack them, a small cup wouldn't stack on top of a larger one without spilling coffee. So if you only moved one cup at a time (rule 1), the physical nature of the cups meant that the only one you could move would be the one at the top of the pile, which had to be the largest.

In contrast to the Coffee Cups puzzle, where only one rule was needed, in the Oranges puzzle, all three rules were needed: There were no physical constraints to force compliance with the rules. The original Tower of Hanoi puzzle was rephrased as "the Doughnuts puzzle," with a waiter who had to deliver three different-size donuts to three customers, following exactly the same rules as for the coffee cups puzzle, but changing the words *cup of coffee* to *doughnut.* Each plate had a peg, and the donuts had to be placed on the peg in the manner shown in Figure 4.1. The doughnuts problem only needs two rules, 1 and 2, because the physical constraints of the pegs force compliance with rule 3.

So now we have three problems, all formally identical. One, the Oranges puzzle, requires three written rules. Another, the Doughnuts puzzle, requires two written rules. The third, the Coffee Cups puzzle, only requires one. The less need for rules, the easier the problem: The Coffee Cups puzzle was easiest (done most quickly, with least error), the Doughnuts next, and the Oranges hardest. Why does it matter so much whether the rules were written or were also incorporated into the physical structure of the puzzle?

External representations add power because the physical structures automatically constrain the actions and interpretations, even though all three rules apply to all the puzzles. Someone programming a computer to solve the task would find all three puzzles to be of equal difficulty and would use the same algorithm to solve all of them. This is because the computer would be unable to take advantage of the physical structures.

Why is the physical form so important to people? Zhang pointed out that the problem really had to be represented in three different ways: first, internally within the problem solver's mind; second, externally in the physical puzzle itself, where the physical constraints play a major role; and finally, once in the mind of the scientists, who study how people solve the problem. It is the scientist who constructs an abstract representation of the problem and the possible moves toward its solution. To the scientist (or computer programmer), all three problems are the same. But to the person who can make use of the physical structures of the puzzle, the more information present in the environment, the less information needs to be maintained within the mind. As a result, for people, the three tasks are very different. In fact, people often don't recognize them as the same problem.

The Coffee Cups and Oranges puzzles may seem peculiar, but they serve as powerful demonstrations of how external representations not only aid in memory and computation but can dramatically affect the way a problem is viewed and the ease with which it can be solved.

FITTING THE REPRESENTATION TO THE TASK

Chapter 3 introduced us to the power of representation. There we discovered that the game of 15 is harder for people than ticktacktoe and that it is easier for people to determine that 284 is less than 912 than to determine that it is less than 312. Logically, the game of 15 and ticktacktoe are the same. Logically, one can determine whether 284 is less than the other alternatives simply by comparing their far left digits, so the judgments ought to be of equal difficulty. These examples make it clear that we do not operate by mathematical or symbolic logic: We operate by perceptual routines. We are especially good at making perceptual judgments, not so good at abstract or symbolic ones.

Saying that we are perceptual creatures does not, however, describe how we operate. Our perceptions are complex, not always operating the way that intuition or commonsense, folk psychology would predict. It is tempting to associate physical variables with the psychological experience. After all, making a light or sound more intense increases its brightness or loudness. Changing the frequency of a light or sound changes its hue or pitch. But to make this simple association would be wrong. The relationship between the physical and psychological dimensions is very complex.

How complex? Well, consider how bright something looks. The more light, the brighter the object, right? Wrong. Look at the following picture:

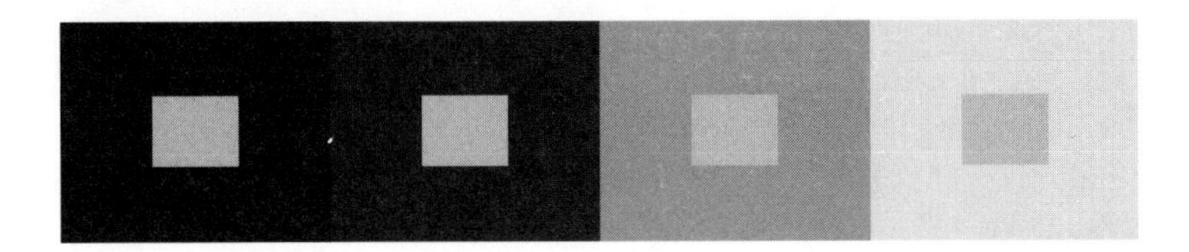

The four inner rectangles all reflect the same amount of light, even though the ones on the left look lighter than the ones on the right. This is because brightness is not the same as light intensity. Brightness depends upon the contrast between an image and surrounding images: A gray patch becomes black when it is next to something bright. Does that sound wrong? Add light to make something darker? Try it. Take this book into a dark closet and then slowly open the door a tiny amount to let some light gradually enter the closet. The rightmost inner square will get darker as you allow more light into the closet.

This principle is exploited in your television set. When the set is off, the screen is gray. Now turn the screen on and note that the image contains black areas and lines. The television set is incapable of placing black on the screen. The screen of a TV set can only emit light: Electrons hit the screen, causing the phosphors on the screen to glow red, green, or blue. A phosphor can only increase the amount of light coming from the screen, not decrease it. The black areas of the screen are created by not sending any electrons to those areas, so the amount of light from a black portion of the screen is exactly the amount that would be emitted if the set were turned off. As we have noted, the screen of the turned-off TV set looks gray, whereas in the picture, we can perceive black. How come? For the same reason that the rightmost inner square looks darker than the leftmost one: It is the contrast that matters.

All this is simply to prove that what you perceive is not necessarily what is there. The *psychology* of perception is very different from the *physics* of perception. Even in cases where more intense lights or sounds are brighter or louder, the relationship is not linear.

Double the amount of light in a room and things do not get very much brighter. This is easy to test. Illuminate a room with a single light bulb and then turn on a second bulb of equal power. You will not notice much difference in the brightness of the room.

Psychologists have determined that perceived intensity of light and sound follows roughly a cube-root law: Brightness and loudness are roughly proportional to the cube root of intensity (the actual law is that brightness or loudness is proportional to $I^{0.3}$).* This means that doubling intensity only makes the perception increase by 20 percent ($2^{0.3} = 1.2$). You have to increase the intensity ten times to make brightness or loudness double.

Most people are completely unaware of this relationship between sound intensity and loudness. Few people realize that there is any difference between what is measured physically and what is perceived psychologically. Even sound engineers, who measure sound logarithmically in decibels (dB), often do not realize that decibels are not an appropriate measure for psychological loudness. A ten times decrease in sound energy means that the number of decibels decreases by ten (−10 dB), but it halves loudness; a one hundred times decrease in sound energy decreases the decibel level by twenty (−20 dB) but makes the sound a quarter as loud; and a thousand times decrease (−30 dB) makes it an eighth as loud.

The complex relationship between sound energy and loudness is often taken advantage of by those who wish to overstate the effects of their manipulation of sound levels. If someone claims "We cut noise levels in half!" be wary: They probably cut noise *intensity* in half, which means that the resulting reduction in *loudness* is barely perceivable to the average listener.

Consider the loudness control of a radio. Here the amount you turn the knob controls sound intensity, which, in turn, affects loudness. Engineers quickly learned that the control couldn't simply control intensity, for if it did, decreasing the control from maximum to the halfway point would hardly make a difference in loudness. Engineers assumed that loudness was proportional to the logarithm of sound intensity, so they made loudness controls logarithmic. That turns out to be wrong, but at least it is better than a linear control. Alas, the engineers who designed light controls never did learn their lessons, which is why it is sometimes hard to control room lighting or, worse, the light of an alarm clock or a clock radio. What the engineers didn't realize about human vision is that it is sensitive to light energy over a range of more than 100 billion to 1 and, moreover, that after being in the dark for half an hour, the eyes become "dark adapted," considerably increasing their sensitivity to light. This is why many clock radios that use a dim light to show the time at night have trouble getting it right: When you first turn off the lights upon going to bed, the dim light of the clock radio might be barely perceptible, but in the middle of the night, after your eyes have adapted through several hours of darkness, the same light may be bright enough to annoy. Why does all this matter? Because it makes a big difference in how information is presented to us by the surface representations of artifacts.

Graphic Representations

Graphs and graphic representations are a surprisingly recent invention. You would think that the perceptual qualities of graphs would have been immediately apparent as a superior method of presenting numerical information. Nope. Graphic presentations were not used by American businesses until the late 1800s and early 1900s, and even then they were sometimes greeted with skepticism. Today they are widely appreciated for their abilities to present trends and comparisons more effectively than by words or lists of numbers. Of course, after something proves effective, it often becomes overused and applied to everything, even where not appropriate. Worse, there are a wide variety of graphic procedures, and not every procedure is appropriate to every situation.

Today it is very easy to create graphs. Too easy. Software programs for the personal computer abound. There are special programs devoted to the construction and analysis of graphically presented data as well as graph-making components in a wide variety of other programs, from spreadsheets to slide-preparation programs to word processors. The problem is that these programs will graph anything, following their own internal rules of logic. Whether those rules also apply to the data and the intended use is a different question, one the programs completely ignore.

My concern is with the violation of psychological principles, whereby graphs are used in inappropriate ways, sometimes deliberately to confuse but more often out of sheer ignorance. My frequent complaints to friends, colleagues, students, and newspapers are commonly met with the excuse "My computer program did that for me automatically; I had no choice." Poor reason: Ignorance of the law is not a valid excuse, whether it be a governmental law or a psychological one.

Let me begin with one of the more common errors in the presentation of information. Suppose that we represent the cities of the world on a map, with the area of the circle that represents each city proportional to population. How well can you judge relative city sizes this way? This is, after all, a typical scheme used in newspaper maps. Well, here are two cities, Beijing and Tokyo, with their areas proportional to their projected size for the year 2000:

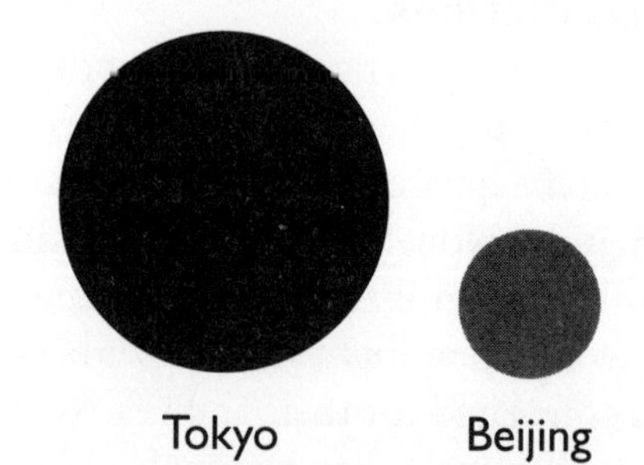

Clearly, Tokyo is expected to be much larger than Beijing, but by how much? Let's try again. Here are the same two cities in a "pie graph," where, once again, the area of the graph represents population:

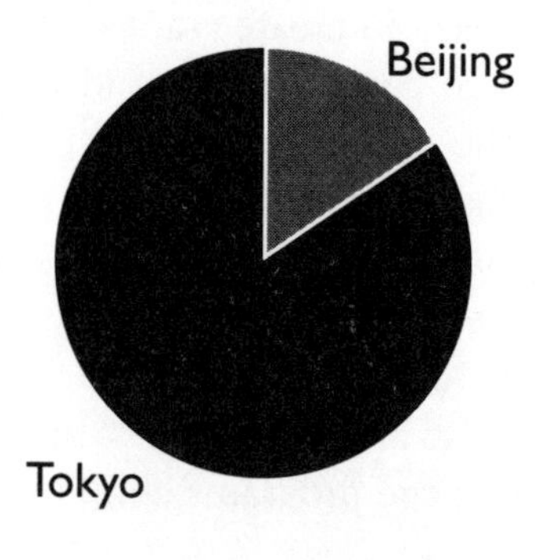

Notice that in the pie graph, the relative dominance of Tokyo over Beijing looks many times larger than it does with the circles. Now compare the same data in bar-graph form, this time where the length of the bar is proportional to the population:

What is the actual ratio of population sizes? In the year 2000, it is estimated that Tokyo will be five times as large as Beijing (30 million versus 6 million estimated population).

Yes, we are perceptual animals, but we excel in seeing patterns, not in forming accurate numerical comparisons from those patterns. Our estimates are most accurate for line lengths—the bar graph—because judgments of line lengths are reasonably accurate, increasing linearly with the physical length of the line. Not so with most things, however, as we have already discussed for loudness and brightness, and as we have experienced with the pie and circle representations.

These distortions of perception are well known by professionals who make graphs. Unfortunately, this knowledge can also be used to mislead the viewer. Inappropriate representations can easily confuse, whether by deliberately making information difficult to discover or by deliberately emphasizing desirable features and obscuring undesirable ones. Figure 4.2 shows how inappropriate use of representational formats can yield misleading results.

The advertisement is intended to show that although the retail price of the Mercedes 300–class automobile was higher than the competition, the cost of ownership over a period of five years was much higher for the competition. How much higher? Well, if you look at the graph, the height of the competitor's bar is 2.5 times that of Mercedes: 250 percent more expensive! Well, no. If you look at the dollar amounts, note that the graph starts out at $42,000: The zero point is far away. The real ratio is 1.15 to 1: only a 15 percent difference in cost. Once the zero point has been taken away from a bar graph, the natural comparison of the line lengths is

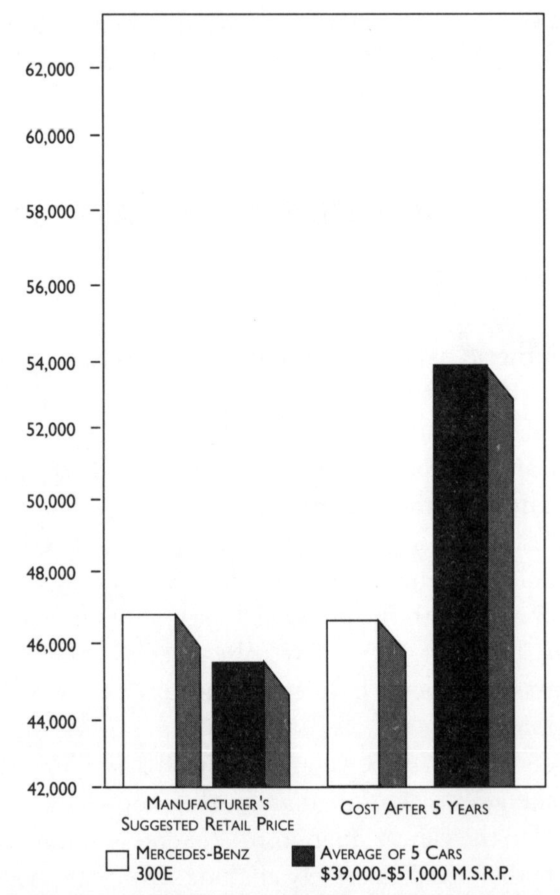

Figure 4.2 Deceptive graphing. Here, the black bar in the right part of the diagram is 2.5 times as high as the white one, implying that cost of ownership is 250% higher than that for Mercedes-Benz cars. But, because the zero point has been moved, the bar graphs are only showing interval scale data and comparison of the ratio of their lengths is inappropriate. The correct ratio is 1.15 to 1: only a 15% difference. (From the *Los Angeles Times, San Diego Edition* [May 21, 1991], p. D5.)

meaningless. True, the correct information is available, and in the technical sense, the graph is accurate. But visual information dominates the initial impression, and not every reader will take the time to do the more difficult, reflective analysis of the numerical information. The dominance of the natural perceptual interpretation is commonly exploited by the advertising industry.

Psychological Scales and Representation

There are numerous psychological principles that can be used to guide the construction of appropriate graphic relationships.* Graphs are interesting beasts, a combination of quantitative, numerical information and qualitative, pictorial information. In fact, graphs are useful because they can translate the abstract, difficult-to-interpret numerical relationships into perceptual, readily visible pictorial ones. But putting everything into pictures is not necessarily good. The left side of Figure 4.3 shows an example of a graph that would be better as the table shown in the right half of the figure. The problem is that the line lengths imply numerical value. When we look at the graph, we are tempted to judge that a Saab (from Sweden) is three times better than a Mercedes-Benz (from Germany) because its line is three times longer.

These examples illustrate that representations should reflect the appropriate power of the medium.

> **Appropriateness principle:** The representation used by the artifact should provide exactly the information acceptable to the task: neither more nor less.

Note that there is no "correct" way to display any particular relationship, but there are definitely incorrect ways. Remember that a representation should support both organization and search and that what is most appropriate depends upon the task to be done. The tabular representation on the right of Figure 4.3 is organized around country of origin, which makes it work well when the user's task is to start with the country and end up with the automobiles. But if the goal is to start with automobiles and determine the country, this particular tabular representation gives no assistance to the search task. It is not a good representation for looking up the automobile name.

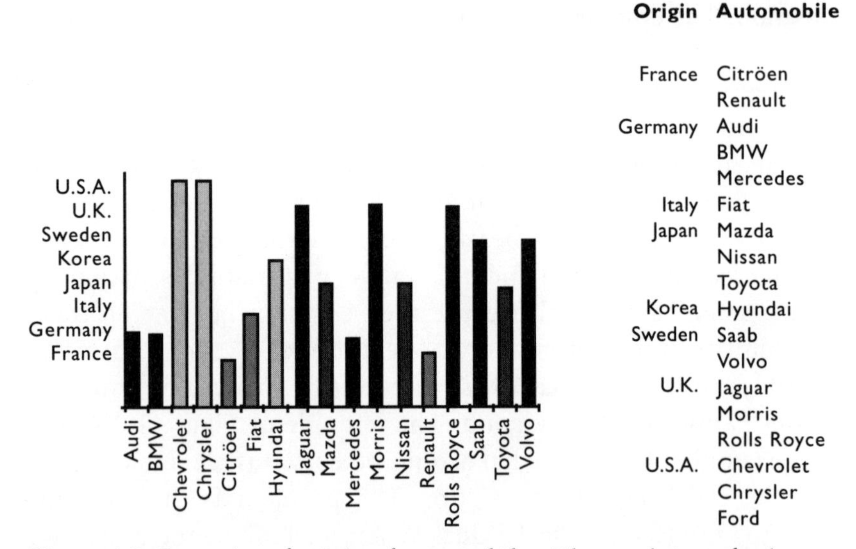

Country of Origin	Brand of Automobile
France	Citröen
	Renault
Germany	Audi
	BMW
	Mercedes
Italy	Fiat
Japan	Mazda
	Nissan
	Toyota
Korea	Hyundai
Sweden	Saab
	Volvo
U.K.	Jaguar
	Morris
	Rolls Royce
U.S.A.	Chevrolet
	Chrysler
	Ford

Figure 4.3 Countries of origin of automobiles. The graph is perfectly accurate, but the use of ratio-scale line lengths to display nominal-scale information is misleading and disturbing. Most people find this graph absurd. The tabular format, shown on the right, is far more appropriate for this type of information. (The graph on the left was inspired by the work of Mackinlay [1986] who deliberately invented this example to demonstrate its futility.)

How could the display be improved? If it were known that the user would always start with the name of a country, then the table is probably near optimum. If it were known that the user would always start with the name of an automobile, then we would alphabetize the automobile list and display country of origin alongside each name. What if both search directions might be used? Here we need one of several things. One method would be a matrix organization: an alphabetical list of automobiles along one axis, an alphabetical list of countries along the other, with a check mark at relevant intersections. Or we might use a network structure, with two lists—one of automobiles, one of countries—linking together the relevant country-automobile pairs. Even here there are tradeoffs in the display: Some are more aesthetically pleasing than others,

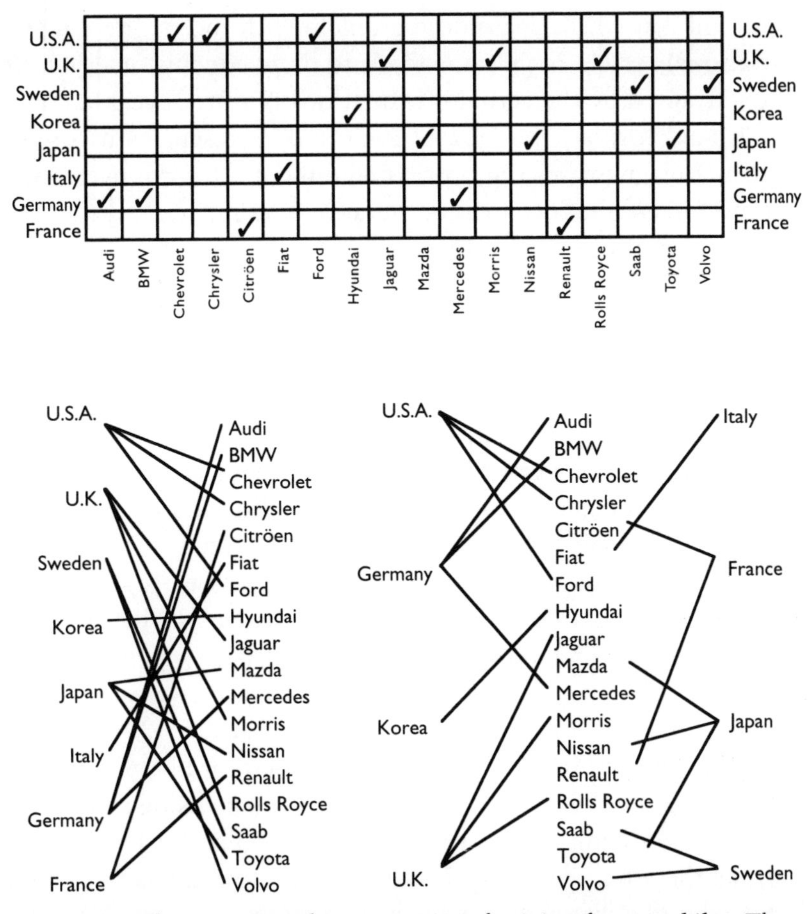

Figure 4.4 Three ways to show countries of origin of automobiles. Three ways of listing automobile manufacturer and country of origin in a manner that allows for symmetrical search: just as easy to go from automobile name to country as from country to automobile name. The matrix is modeled on the pill matrix of Chapter 3. However, here it is difficult to follow the long rows and columns. The association diagram in the bottom left is cluttered and hard to follow. The association diagram in the bottom right is easiest to use, except that the country names are no longer in alphabetical order and therefore more difficult to find.

some work more efficiently than others. Compare the several possibilities in Figure 4.4.

Figure 4.3 shows what happens when too powerful a representation is used. Basically, the problem with too strong a representational system is that one tends to draw conclusions that are not warranted by the actual information. With too weak a representational format, the lack of a natural representation for the information increases the processing difficulty for the user, forcing more mental effort and computation and reducing the ability of representations to exploit the power of human perceptual processing.

Digital Versus Analog Displays

It is rarely possible to discuss representational formats without someone complaining about digital watches. "What a miserable device," the complaint goes. "Analog watches are far superior. You can tell at a glance roughly what time it is and how much is left to go. You can't do that with a digital watch." Proponents of analog watches will tell you that people wearing a digital watch can always be detected, because when asked the time, they will invariably reply with excess precision, as in "8:42," whereas "8:40" would do or, better yet, "about twenty minutes to nine."

The same argument, by the way, can be heard between those who like automobile speed to be indicated by an analog display—a moving hand along a dial face—and those who prefer a digital display. Is the whole world to be divided into two kinds of people, those who like digital versus those who don't?

Which representation is superior? Neither: It all depends upon the task. Some representations are superior for some tasks, others for others. Let's consider the automobile speed indicator: the speedometer. Which is better, a digital or an analog display?

Answer: What do you want to know? If you want to know the exact numerical value of the speed, then digital is superior because it presents the answer in precisely the form in which you need to know it: a number. The answer is easy to read rapidly and accurately. With an analog dial, the exact value of speed is more difficult to determine: The pointer has to be interpreted to yield the exact value.

Most of the time, however, we do not need to know the exact

numerical value of our speed: Either we need a rough estimation, or we need to know whether we are above or below some critical speed, such as the legal speed limit. Here the analog speedometer is superior, but only if the dial contains a highly visible reference mark. Then you can simply look to see on which side of the mark the pointer is: Above and you are going too fast, below and you are going slower than the limit. Moreover, the distance between the pointer and the mark, as well as the speed and direction of the pointer movement, all convey useful information about approximate speed and rate of acceleration. In a similar fashion, after some experience with the speedometer, the pointer angles take on meaningful values, and the angle can be determined accurately enough for most purposes with a rapid glance, yielding an approximate value of speed.

In commercial airplanes, pilots are provided with "speed bugs," little reference marks that they can move to critical spots along the airspeed indicator. This makes it easier to transform their reading of the analog dial position into "more than" or "less than" judgments of airspeed. Analog meters are clearly superior for this purpose.

In the automobile, the advantages of the analog readout are mixed because of the lack of a flexible way of marking the critical speed of interest. As a result, neither analog nor digital speedometers seem superior. If the critical speed is 35, then with the analog dial, one must figure out where 35 might lie on the scale, then whether the pointer is above or below it. With a digital speedometer, one must mentally subtract 35 from the indicated number to determine whether one is above or below. Neither method seems particularly virtuous.

What is the best method? Why choose? Why not have both? Why not have an analog speed indicator with a digital display alongside? Today's commercial aircraft use combination analog and digital displays for airspeed and altitude, with easy ways to set "bugs" at critical values. A similar philosophy might be effective in the automobile.

Altitude meters in aircraft used to have three hands, but pilots made so many errors in reading them that today most altimeters have replaced two of the hands with a digital readout, retaining the analog pointer for only the least significant digits. Thus if the airplane's altitude is 31,255 feet, the digital meter will display the altitude in thousands of feet as "31," and the analog meter will display the hundreds—namely, "2.55" (255 feet). This dual display has several advantages. The most significant reading—for which accurate values are important—is given digitally, in thousands of feet. The pointer shows hundreds of feet, and for this purpose, precise numerical information is seldom required. Moreover, the pilot can readily determine whether the plane is changing altitude upward or downward simply by seeing which way and how fast the hand is spinning. With the digital readout, if the change in altitude is too fast, all that you can see is a blur. (Same problem with digital speedometers.)

What about the clock? Which is best: digital or analog? The answer is the same: It all depends upon the task. We have to bear in mind, however, that watches suffer a major disadvantage when compared to the speedometer. With the speedometer, there is one pointer, one scale. With watches, there are two or three pointers, two different scales. It is not easy to learn to read time on analog watches: Children have considerable difficulty, and even adults who read the time quickly make errors, confusing the hour hand with the minute hand, and vice versa, so that 3:20 might be read as 4:18. With today's watches and clocks we no longer are forced to choose between analog and digital. It is possible to have both.

FITTING THE REPRESENTATION TO THE PERSON

Did you ever notice how much information is provided for us by the world? Thank goodness, else we would never manage. Look at how many different physical objects we use in our lives: knives, forks, pencils, paper clips, shoes, shoelaces, buttons, zippers—I once estimated that we all are probably familiar with twenty thousand different objects, each small, specialized, and requiring learning. How do we cope? How do we manage to learn about each of these individual items? The answer is that the physical design of an object makes all the difference. You can often tell just by looking at something what function it serves or, at least, which parts you are supposed to hold, push, or pull; which parts operate upon other devices. A thumbtack has an obvious place to push, an obvious point that can be used to pierce, hold, or even lift objects. Most devices provide enough clues to their operation that even though some instruction might still be needed, the instruction can be accomplished in just a few words, or perhaps just by watching someone else use it once. Physical devices have affordances, mappings, and constraints that greatly aid in figuring out how they can be used. Of course, when the device is poorly designed or constructed, these same factors can greatly hinder its use. This is the story I told in *The Design of Everyday Things.*

Physical artifacts can be designed so that they are easy to learn, easy to use. The same is true for cognitive artifacts, although here some new principles must be provided. We need to consider the nature of the task to be performed and the powers of the human.

We humans seek understanding, causes, and purpose. We are good at remembering experiences, good at stories and events, bad with the minutiae of modern life. We are attentive to our surrounds, remarkably quick to notice changes. And we see patterns and meanings even when they are obscure and hidden. These very same characteristics, however, can conflict with the demands of the modern industrial, technological life. The conflicts are made worse by the technology that is imposed upon us on its terms instead of ours. The conflicts could be minimized or even eliminated if the technology came to us on our terms.

Furthermore, we are social creatures. We communicate and work well in small groups, sharing and cooperating to accomplish tasks beyond the capability of the single individual. The cooperation is aided through the communicative powers of language and body: spoken and written words, gestures, eye contact, and facial expressions. People are biologically predisposed to work in rich, ever-changing, sociocultural environments. We exploit any relationships we can find and invent interpretations. All this aids us in making sense of an otherwise chaotic world.

Today we live in an information-based technological world. The problem is that this is an invisible technology. Knowledge and information are invisible. They have no natural form. It is up to the conveyer of the information and knowledge to provide shape, substance, and organization. The irony is that too much of our artificial world is oversimplified, overabstract, thereby taking away our most powerful capabilities.

Information media do not necessarily take on a form amenable to humans. They are true internal artifacts, in that information is abstract and invisible. The information tends to be represented internally in the same manner, regardless of its content. This is especially true of digital media, in which everything gets transformed into a numerical representation expressed as binary digits—a long sequence of 0s and 1s (usually encoded as two levels of electrical charge). More and more of our media use a digital format for storage and transmission—television, telephone, radio, books. Digital signals offer a number of advantages: They are relatively easy to process and work with electronically, they offer great immunity to interference by electrical noise, and they are the natural medium of storage and processing used by computer systems.

Digital media have a number of disadvantages. A major problem is that of making access feasible, understandable. The common format for everything doesn't help: The most beautiful painting, the most stirring music, the most profound thoughts are all reduced to the identical format of internal states of the artifact. In fact, one cannot tell whether a given message is music or art, beautiful or ugly, from the internal representation. The medium is completely neutral with respect to the content. Hence the power of the medium: The signals can be transmitted and operated on without much regard for their content. It is only when people get into the picture that the form and content matter. Humans need a meaningful, accessible representation: sounds, sights, touch, organized in meaningful, interpretable ways. The result, however, is that we are ever more dependent upon the design of our devices to make the information visible and to make the artifact usable.

The new-fashioned information artifacts take on arbitrary shape and form. There is no natural mapping, no natural principles of operation. The critical operations all take place invisibly through internal representations. If we are to be able to use these artifacts easily and efficiently, the designers have to provide us with assistance, with an understandable, coherent structure. We are in the hands of the designers, who have the power to make the artifact meaningful, to provide substance and richness, and to make its use support the activities of interest. The best of the artifacts will become invisible, fitting the task so perfectly that they merge with it. They will be a delight to use.

Design should be like telling a story. The design team should start by considering the tasks that the artifact is intended to serve and the people who will use it. To accomplish this, the design team must include expertise in human cognition, in social interaction, in the task that is to be supported, and in the technologies that will be used. Appropriate design is a hard job. But without it, our tools will continue to frustrate, to confuse more than clarify, and to get in the way rather than merge with the task. The power of information artifacts is that they provide an unrivaled opportunity to enhance our lives. The danger is that they can add to the stress of everyday existence.

TECHNOLOGIES HAVE AFFORDANCES

"You know what bothers me about the difference between television and newspapers?" a friend asked. With television the ads are inescapable. They blast right at you—you can't avoid them, except by leaving the room. It's not that way at all with newspapers. In fact, it's just the opposite problem. Sometimes, when I really want to read the ads, I don't even see them. Why is that?"

The problem lies with the differences between the affordances of television and newspapers. The *affordances* of an object refers to its possible functions: A chair affords support, whether for standing, sitting, or the placement of objects. A pencil affords lifting, grasping, turning, poking, supporting, tapping, and of course, writing. In design, the critical issue is perceived affordances: what people perceive the object can do. We tend to use objects in ways suggested by the most salient perceived affordances, not in ways that are difficult to discover (hence the fact that many owners of electronic devices often fail to use some of their most powerful features—indeed, often do not even know of their existence).

Affordance also applies to technologies. Different technologies afford different operations. That is, they make some things easy to do, others difficult or impossible. It should come as no surprise that those things that the affordances make easy are apt to get done, those things that the affordances make difficult are not apt to get done.

In the case of my friend's complaint about the inescapability of television ads and the invisibility of newspaper ads, the culprit was the differences between the affordances of the serial, time-paced presentation of television versus the parallel, self-paced presentation of the printed page coupled with the fact that, on the whole, people can only attend to one thing at a time. With the television set, there is only one thing to look at, one sound channel to listen to. One message at a time.* Sure, we can daydream or look away from the television set, but unless we make some active effort to avoid it, the material impinges upon the brain, upon the conscious mind. Couple this with the cleverness of television advertisers to exploit the seductive experiential quality of the medium and it becomes clear how, as viewers, we are readily hooked.

The printed page provides a very different story. Here, as readers, we guide the intake of information. We must actively move our eyes across the page. Multiple articles and advertisements appear on each page, and because people can only read one thing at a time, we have to actively choose which to read: The act of selecting one automatically excludes the others. Moreover, once we start reading an article, our eyes track its location, even across pages, thereby missing any other material on the page.

A television channel presents different information by devoting different time slots to different information contents, presenting everything in the same spatial location. Print media—such as books, magazines, and newspapers—present different information by devoting different spatial locations to different information contents, presenting everything at the same time. The differences in the use of space and time between the television and print media yield different affordances. One result is that we find it difficult to escape television commercials because we focus on spatial locations. Even if we divert our attention as soon as we become aware of the commercial, it is too late. In a newspaper, the advertisements are not in the same spatial location as the stories, so the eyes can miss them, even when we would prefer not to.

Television organizes its information in time, newspapers in space. The result is that television paces the reader (it is event-paced), whereas with the printed page, it is the reader who sets the pace (it is self-paced). This is why the printed page provides better affordances for reflection than does the television show. Because reading is self-paced, there is time to pause and reflect upon what has been read, thus performing a deeper analysis than is possible with the event-paced affordances of television.

When a technology attempts to force a medium into a usage that violates its affordances, then the medium gets in the way. The result makes the difference between a humane technology and an inhumane one. Let me give an example of a reasonable idea made inhumane by the affordances of the medium: voice-messaging systems.

The phrase *voice-messaging systems* refers to several different kinds of things. The goal is to make it easy to provide customers with the specific information they need through selective messaging. These systems have proliferated. Companies find them attractive. As for the callers? Annoyed and upset. Furious even. Why? The concept is reasonable; the failure lies in the affordances. The version that upsets users so much does so because it violates the affordances of the telephone medium.

There are a number of problems from the point of view of the caller, but most are traceable to the fact that the technology available to the caller is tiny and impoverished: It is a poor vehicle for communication. All the caller has is a keypad with twelve buttons and a telephone instrument that allows voice messages to be spoken and listened to. Somehow, using these minimal tools, the caller has to get access to just the correct information out of all the thousands of possibilities offered by the company. The system gives explanations and alternative choices through voice messages to the caller: The limited affordances of voice in this context are at the heart of the problem.

Voice is a serial medium of relatively low-speed, low-capacity communication. It is transient: The information is available only for the duration of the sound itself. Therefore, if a series of alternatives is to be presented to the listener, they have to be limited in number to the amount that can be retained in working memory: Ten would be too much, five would be acceptable if the caller could pay full attention to the message and not be distracted. Three would be safe. Three alternatives? Where the company might have perhaps five hundred different messages and locations to which the call could be directed? Even limiting the alternatives to five would create problems.

The transient nature of voice is a major problem that fundamentally limits the service. The speed of speaking is another. Each alternative might take one or two seconds to speak. A message with three alternatives would thereby take as long as six seconds; ten alternatives might take twenty seconds. That's a long time. There are two standard solutions to these problems, usually used together.

The first solution is to try to present a larger number of alternatives but encourage the user to punch in the number of the desired one as soon as it is heard. This avoids the working memory problem: Listen to each, make a yes/no decision, and then go on to the next. This, of course, assumes that you can recognize the one you want when you hear it. In fact, often the only way to tell which one you want is to listen to all the alternatives and make your selection by a combination of knowing some are irrelevant, some seem relevant.

The second solution is to make the selection process hierarchical. Each set of choices leads to another set of choices, until the end point is reached. If there are n alternatives the first time, and if each of the n choices leads to a second level, also with n choices, then the two levels combined let you reach n^2 alternatives. With three levels, one could reach n^3 alternatives, and so on. If there are five alternatives per level and five hundred possible destinations, four levels are required. Four levels, each with five alternatives? It might take ten seconds at each level to listen to the choices and select one, so it would take forty seconds of listening and keypunching to get to the desired location: if, indeed, the end point turned out to be the desired one. Forty seconds is a long time. Try it: Stop right now and do nothing for forty seconds.

The telephone system lacks some critical affordances—in particular, those required for graceful error correction. Suppose you push the wrong number, either by accident or because you had misinterpreted the message. How would you correct it? In the first case, you would probably realize the error as soon as you made it. The system could make one of its alternatives a chance to go back to the previous level, but if it did so, it would either have to increase the number of alternatives at each level (in my example, to six) or replace one of the other alternatives. If the first procedure is used, the time per level increases to twelve seconds, and the total time increases to forty-eight seconds (along with the chance of overloading working memory). If the second procedure is used, then there are only effectively four alternatives per level, so it would take five levels instead of four to reach the destination: The time would increase to fifty seconds.

Let me illustrate the problems with an example. I subscribe to the *Pocket Edition of the Official Airline Guide* (OAG), a monthly publication small enough to carry in a shirt pocket. American Airlines has an automated telephone flight information system ("Dial-AA-Flight"), so I decided to compare how long it would take to find some American Airlines flights using its automated system with the time required using the OAG pocket guide.

The task I set for myself was to see how long it would take to find all American Airlines flights between San Diego, California, and Detroit, Michigan. First, I used the pocket guide: Forty seconds and I had the answer. Then I dialed the phone number of the American Airlines system. I was greeted by some music and voice instructions. One of the possibilities was a training session, so I selected it, and when it was finished, I hung up so as not to bias the timing with training time. Then I called again and this time chose the option for experienced folks. That got me to a second level. The system was well done. American Airlines only presented two or three alternatives at each level in an attempt to keep the information within working-memory span. I had no trouble remembering the alternatives, so after selecting flight-scheduling and fare information, I specified my departure and destination cities by typing the first four letters on the telephone keypad. Each digit on the telephone has three letters, so the digits were ambiguous. (For example, both San Diego and San Francisco are specified by the same digit sequence: 7263.) At the next level, I had to select among the resulting ambiguities. At the eighth level I was asked whether I wanted the flight to leave in AM or PM. That was strange: What if I didn't know when I wanted to leave? What I wanted to do was to compare the alternatives. I had no option but to select one. When going east from California I almost always leave in the morning, so I typed "1*" for AM.

On the ninth level, I was asked to type in the time of the flight that I was interested in. Huh? I didn't know when the flights left, that's why I was calling—so that I might find out. I wasn't sure what number to type, so I decided to use the earliest time I would be willing to leave: 8:00 AM. Alas, I made an error—I typed "7" instead of "8." How could I correct the error? I didn't know. I had not been told. I cleverly decided to type an illegal time, assuming that this would get a message saying something like "That time was not proper, please try again." I entered the time "777*" and waited. Instead of a friendly error message, I got an even friendlier one saying that a human was on the way to help me. I hung up. Total time, not counting training, 128 seconds. Two minutes and eight seconds—over three times longer than with the book—and the book gave me all the flights, but with the phone system, I had gotten none of them.

I still don't know whether it is possible to get all the flights with the phone system, unless you do it one at a time (typing in each possible departure time?). Yet I never make a reservation by simply saying what day and time I wish to leave: I always compare the range of possibilities and, after some reflection, select the one that best fits my needs for that trip. I think I will stick with my travel agent and the OAG.

No wonder customers are irate. The idea of delivering information by telephone is actually a pretty good one, but the medium does not support it. Voice is too slow, too transient, and the telephone keypad too limited. Voice-messaging systems will only work painlessly if the medium is transformed to have better affordances, probably by changing the equipment available to the caller. Notice how the printed medium of the pocket OAG is superior for rapid scanning and efficient presentation of information. Voice simply won't do: It has the wrong affordances. Voice is serial. Vision is parallel. Voice is transient. Printed or displayed images are relatively permanent. If my telephone had a high-quality visual display with high resolution and contrast, then the voice-message system could be replaced with a visual-message system. I would always see a page of information with multiple alternatives, including simple ways to get to human assistance at any point. It would not be hard to present twenty to fifty alternatives on a properly configured visual display, so five hundred alternatives could be reached in two to three levels. Reading is faster than listening, and if the displays were properly designed to allow rapid scanning and easy error correction (a very big "if," given the poor track record of existing technology), the affordances of visual presentation would transform a clumsy, inefficient scheme into a workable one.

Suppose that the visually presented system were in use, would this solve the problems? Here I am, foisting yet another technological solution on us, yet another way to avoid having people talk to people. My answer here is the same as always: There is no standard answer, it all depends upon the situation. Sometimes it is better for all concerned to get rapid, efficient access to the answer, even if technologically presented. Sometimes we need human interaction. A well-designed system will provide both.

The affordances of the medium do make a difference. My simple analyses and test show some of the reasons these voice-message systems are so disliked. But this raises another question: If they are so universally disliked, why are these systems used? Why are their numbers increasing? Why this effort to force the technology to do something for which it is so ill suited? The answer is that the systems appear to provide major benefits to the company. They relieve employees from a continual barrage of phone callers who have standard questions, and they do save the cost of numerous telephone and information operators. This attitude, of course, neglects the cost to the callers who are frustrated and angered by the system. One company officer who ordered the machines taken out of his company described it to his employees this way: "What you're saying is your time is worth more than their time."

Voice-mail and voice-messaging systems do have situations where they work well. They often provide a superior means of delivering personal messages. One person told me of a system for providing information about films and theater schedules that he considered a superior use of the technology—better than any other existing method. (I have not had a chance to try it.) The technology can work well in appropriate situations for appropriate tasks.

Technology usually provides a series of tradeoffs. Each asset is offset by a deficit. It is always necessary to decide whether the assets outweigh the deficits. Frequently, the tradeoffs fall differently upon different people. A major problem occurs when those who suffer from technology's deficits and those who benefit are not the same people. In the telephone system, the benefits are to the company, the burdens fall upon the users. This kind of tradeoff I have come to call Grudin's law, after Jonathan Grudin, who first proposed it:

> **Grudin's law:** When those who benefit are not those who do the work, then the technology is likely to fail or, at least, be subverted.

Grudin's law very definitely applies to the voice-messaging systems. May they die a rapid death.

FIVE

THE HUMAN MIND

NOT MANY YEARS AGO, MANY SCIENTISTS (INCLUDING ME) THOUGHT they were on the verge of a great scientific breakthrough: They could create artificially intelligent machines. Advances in our understanding of information theory, human problem solving, and language enabled them to develop machines with limited powers to manipulate information, answer questions, and solve problems. It wouldn't be long, respected scientists prophesied, before machines would equal or even surpass human intelligence.

Ah, science. The point to remember while listening to scientists is that they are simultaneously very conservative and extremely radical. Scientists are trained to be skeptics, to distrust any new result or pronouncement, and to doubt the initial reports of any new phenomenon or observation. On the other hand, once scientists believe those findings, then all bets are off—they believe they have discovered the key to the whole problem: A few more years, and yet another secret of the universe will have yielded.

Have we managed to record the electrical potentials from a living, normal brain cell? Hurrah, we are close to a full understanding of the brain! Do we understand how billiard balls collide and rebound? Hurrah, we are close to understanding the motions of the universe! Can we make a machine that can play some games better than people, prove some mathematical theorems, and understand simple sentences? Hurrah, we are ready to replicate human thought!

The overconfidence of scientists is probably necessary: Let them believe they are on the trail of something big, something important, and they will slave away for their entire lives, working, arguing, debating, exploring. The self-confidence keeps them going, keeps them excited and energetic about their work. Take away the faith, and morale sags, the work languishes. Self-confidence is a necessary human state. Faith in oneself. Faith in one's activities. People are very good at believing in what they do, even in the face of contrary evidence, in the face of others who believe the opposite. Strong beliefs are important to human activity: Why else would we do so many of the things that we do?

Scientific progress may occur because of the emotional beliefs and attachments of people, but the irony is that the science of human cognition ignores the very things that make human progress so exciting and interesting. Instead, the focus is on things that can be studied, measured, and reproduced in carefully controlled laboratory experiments. This leaves out much of social interaction, humor, emotion, motivation, and creativity. It includes psychophysics (the sensitivity of the sensory systems), reaction time (how quickly people respond to flashing lights and sounds), memory (for lists of words and carefully crafted stories), and problem solving (with small, well-defined puzzles). The mind is conceived as a powerful information processor. In comes information from the eyes and ears, out come movement and language. In between, incoming sensory information is transformed into an electrochemical representation.

This hard approach to science has its virtues. A lot has been learned about human cognition. The problem is, there is much more to human cognition than that. Humans are immensely complicated creatures. The brain is an exquisitely complex device, tiny in size, considering that it contains some 10^{12} separate nerve cells, each connecting to an average of 10,000 other cells: some 10^{16} connections in all. If each nerve cell passes an electrical impulse only ten times a second, that's 10^{17} impulses a second. That number—a 1 followed by seventeen 0s—is unimaginably large. Moreover, that's not all that is going on. The brain exudes all sorts of liquids. Chemicals bathe the nerve cells, hormones squirt here and there. It is no wonder that the magazine *Scientific American* has characterized the human brain as "the most complex structure in the known universe."

There is far more to human cognition than what goes on in the brain: we are social, interacting creatures. Many of the important parts of life go on outside the head, in our interactions with the world, in our interactions with each other. Just as an important part of science is the value system that keeps the scientist motivated for years on end pursuing what might turn out to be a fruitless search in a promising but eventually inappropriate direction, many of the important parts of human activity come about through our social interactions and shared knowledge and beliefs, not just the activity within the individual head.

We have a long way to go before we can hope to make machines with the same abilities as people. You can think of this statement as either optimistic or pessimistic, depending upon your point of view, but the differences between humans and machines are greater than their similarities. On the whole, I consider this to be good: It means that we have a chance of complementing our abilities. Machines tend to operate by quite different principles than the human brain, so the powers and weaknesses of machines are very different from those of people. As a result, the two together—the powers of the machine and the powers of the person—complement one other, leading to the possibility that the combination will be more fruitful and powerful than either alone. If we design things properly, that is.

Machines—and scientific reasoning—tend to be logical and consistent. It is no accident that the artificial language of arithmetic and mathematics is ideally suited for machines, whereas the natural language of people is not. Artificial languages tend to have a nice, formal structure. There are rules to be followed, and years of practice and education are required to get people comfortable with and proficient in the rules and structures of mathematics. Numerical operations do not come naturally to people, and not only must considerable training be provided, but even with the aid of external artifacts—paper and pencil, tables of numerical properties, calculators, computers—it is easy to make mistakes. Machines are quite competent at arithmetic and rules: They only make mistakes when they physically break down.

Human beings tend to be driven by patterns, by events. People are highly emotional as well. We empathize with the problems of others, sympathize, or criticize. We pass judgments, find reasons, try to interpret and understand. Human language is enormously complex by mathematical standards: Many of its properties still defy scientific description. Nonetheless, all normal humans learn their native language effortlessly, without schooling, without any formal instruction.

It shouldn't come as any great surprise that the human brain works quite differently than a mechanical one: After all, they operate by very different principles, and even the most complex computer is puny compared to the power of the brain. But what about the difference between people and animals? After all, aren't the brains of humans and apes very similar?

The great superiority of human reasoning over that of the apes has long puzzled students of cognition, because from a biological point of view, the ape family (which includes the gorilla and the chimpanzee, among others) is extremely similar to the human's. Nonetheless, humans have language, can reason, and develop cumulative knowledge bases and large, cooperative societies. Humans construct technology. Actually, apes do all of this, but to a very limited degree. Their tools are relatively simple, such as pruned branches used to fish termites from nests or pairs of stones used to smash open nuts. Their cooperative strategies are limited, and their natural use of language is extremely limited. Even apes that have been taught language in various scientific experiments are very limited in the extent to which they can use this language in novel, constructive ways.

The human brain is not just an ape's brain with some extra stuff added on. Evolution proceeds by modifying and adding to existing structures, and although we are close relatives of apes, we are not descended from them: Rather, apes and humans both descended from a common ancestor some 6 to 8 million years ago. The result is that our brain differs from that of the ape, through both modification and addition. Our vocal apparatus is different, thereby allowing us to produce the rapid, fluent sounds of language. Our motor skills are more agile, with better timing and coordination. And our brains have powers that are absent in the ape. Most important, the human is capable of reflective thought.

Consider some of the intellectual abilities that are unique to humans:

Art

Games and sports

Humor and jokes

Language, both its invention and its creative, constructive use

Music

Ritual

Satire, whether by mime, picture, or words

Schooling, and thereby cumulative knowledge and culture, in which the discoveries of one generation are passed on to the next, so that new achievements build upon prior developments

Storytelling

The appreciation of beauty

These are considerable skills and abilities, and their absence from animal and machine intelligence is significant. These abilities all require a sophisticated mind, one capable of representing first knowledge and then metaknowledge, the ability to form representations, to compare one representation with another, and to be able to form causal explanations of the events of the world. We represent the needs, motives, desires, and capabilities of ourselves and others. The power of our mind is in its representational capacity, including its ability to take another's point of view. We attribute the actions of others to their intentions and desires, which means we have a theory of our own mind and that of others. We are explanatory creatures: We develop explanations for the events of the world and for the actions of people, both ourselves and others. We find reasons, causes, explanations. All of these require us to create "mental models," mental scenarios in which we construct representative explanatory descriptions. Mental models allow us to understand prior experiences, the better to predict future ones. Mental models also give us guidance and assistance in knowing what to expect and how to respond in novel or dangerous situations.

Modern studies of evolution, animal intelligence, the capabilities of the developing human infant, and the representational abilities of people are starting to converge upon a cohesive, consistent understanding of the factors that led to the development of human intellect. Several lines of evidence converge:

- *Social interaction:* Only the most capable of animals can deceive properly, for deceit requires knowing what action will cause the other animal to be fooled; in other words, it requires knowledge of—a representation of—the other animal's knowledge.

  Similarly, only the most advanced animals are capable of true, cooperative social behavior. Sure, insects are social animals, working together to enhance the survival of the group. But this is unplanned, unintentional. Our cooperation is intentional and planned. We even devise governmental and legal structures to ensure its continuance. The result is that most of us live in complex, cooperative societies, where the combined efforts of the group far exceed the capabilities of any individual. Yes, there are rivalries and feuds among individuals and, at times, deadly wars between competing societies, but these do not change the fact that our social cooperative efforts are an essential part of human civilization. This is how we weather the elements, get through famine and flood. This is how we develop knowledge and education, formal methods of schooling that allow each generation to benefit from the lessons of the previous generations. We succeed so well at survival that a considerable fraction of our time goes toward the pursuit of enjoyment in the arts, literature, sports, and entertainment. No other animal has such luxury or ability.

- *Teaching:* Other animals do learn through imitation, but that requires skill on the part of the learner, not the teacher. Only humans are skilled at teaching, at understanding how a topic has to be presented, simplified, and broken down into elementary concepts in order for the learner to acquire it most readily and successfully. Only humans have formal instruction, whether in schools, spontaneously formed groups of learners around a teacher, or a period of apprenticeship.
- *Artifacts:* A number of animals use tools, but in limited ways. Animals seem restricted to simple tools as opposed to complex, multicomponent ones. They do not make tools that aid in toolmaking as we do. Our complex tools require specialized skills and materials. And only the human has cognitive artifacts.

THE ORIGIN OF HUMAN INTELLIGENCE

In an important book, the psychologist Mervin Donald argues that human intelligence has evolved through a series of evolutionary steps to its present form—a form, moreover, that is highly dependent upon external, artificial representations for its power. Donald argues that the human is actually three evolutionary steps above the ape. In particular, he argues that the ape has primarily an episodic memory structure, one capable of experiencing complex events but not of making great abstractions from those events.

For Donald, human intelligence evolved in a nice sequential manner. Note that evolution operates by modifying existing structures, usually in small, slow steps. Each generation of animals represents only a slight modification of previous generations. As a result, large changes take tens and hundreds of thousands, or even millions, of years: By evolutionary standards, a thousand years is but an instant, hardly worth noticing. Each change, however, has to be viable. That is, if you want to postulate the chain of evolution that got us from some ancestor, somewhat like the ape, to today, you have to show a large sequence of changes, each one of which is functional, each a sufficient advance from its predecessor that it will survive in the competition among species. In engineering, when one builds a new device by tinkering with the old, modifying this, adding that, the result is called a "kludge," a haphazard, unstructured collection of parts that, nonetheless, functions well. Good scientists and engineers don't believe in kludges: When they need to make a new machine, they start over again. This may take longer, but the result is a device that is well structured, easy to understand, and easy to repair. Evolution doesn't care about the scientific method: Evolution builds kludges.

As previously mentioned, humans did not evolve from apes: Humans and apes had a common ancestor, some 6 million to 8 million years ago, and both have evolved considerably since then. Therefore in postulating an evolutionary sequence, we know where we have ended up—with us—but we do not know where we started from, nor do we know the steps in between. Still, we assume that apes and chimpanzees are more similar to the starting point than we are, so that observing their capabilities gives us evidence about our common ancestor. This is the route Donald takes. He postulates four evolutionary stages in the development of human cognition:

1. *Episodic memory:* This is the cognitive level of apes and, presumably, our common ancestor. At this level of cognition, animals are limited in their ability to form mental representations of events in the world. In particular, they are limited to the ability to remember particular episodes and states of the world, hence the name "the stage of episodic memory." I would have called this the "experiential stage."
2. *Mimesis:* We can observe animals that are in the episodic stage, but now we move to a hypothetical stage of human evolution: the stage of mime—mimesis. Here the ability to form internal representations is expanded to include desires and wants, and this is supplemented by the ability to act out—to mime—these desires. For the first time, sophisticated communication of intentions and mental states becomes possible.

3. *Mythic:* Mime is the first stage leading toward language. The second, still hypothetical, stage of human cognitive evolution is that of language development, where mime has given rise to language. Now, for the first time, animals can communicate rich concepts and thoughts to one another, greatly aiding group planning and activity. With language and increased representational capacity comes the telling of stories, myths, that provide explanations of the events of life.
4. *External representation:* The final stage is that of the modern human. Today our abilities to mime, use language, and reason are expanded through the power of writing, external representations, and tools. In other words, in today's world, we have taken evolution into our own hands, providing external devices—what I have called "cognitive artifacts"—to expand our abilities beyond that which our biological heritage alone makes possible. The future of human evolution is through technology.

Representation plays a major role in these analyses, but except for the last stage, "external representation," not the forms of external representation that we have considered so far. No, a critical factor underlying intelligent brains is the internal representations that the animal has and the nature of the processes that operate upon those representations. What operations can be performed upon the animal's internal representations? How aware is the animal of these representations? The development of self-awareness and consciousness is a critical step in the evolution of human intelligence.

Social interaction demands more of animals than does solitary behavior. A solitary animal only needs to represent the current state of the world, have some experiential record of the past, and be able to make predictions of the future based upon those experiences. Animals that work together must learn to synchronize and coordinate their activities. In highly social animals such as the baboon, a considerable part of each day is spent establishing and maintaining social roles. The most fruitful interaction requires that each animal knows what the other animal knows and plans to do. Of all animals, this ability is most highly developed in humans, next most developed in apes and whales (including dolphins).

Simple synchronization and cooperation do not necessarily require much brainpower. Ants cooperate in an impressive variety of tasks, but this cooperation is built in, part of the wiring of their nervous systems and not a result of any conscious desire to work together. Ants simply respond to the situation they perceive: The apparent cooperation is actually just the result of the automatic patterns formed through individual responsiveness. In similar fashion, the synchronized, orderly patterns made by birds flying in a V-shaped formation or schools of fish darting this way and that do not result from any knowledge of the overall pattern or conscious attempt to form a V or a school, but rather from very simple individual responses to the situation. Thus, if each animal flies so as to keep just slightly behind and on the outside of the others, a V-shaped pattern automatically emerges, despite the fact that no bird knows anything at all about Vs. You might note that this simple behavior doesn't guarantee a V, but flocks don't always form Vs either. Still, it shows how simple, wired-in behaviors can lead to sophisticated behavior. The problem with these built-in cooperative patterns is that they are fixed and unvarying. They do not change even when the conditions requiring cooperation no longer exist. It is this lack of flexibility that distinguishes evolutionary, biological cooperation (the wired-in kind) from intentional, cognitive cooperation.

True cooperative behavior requires some sort of shared knowledge and conscious desire to cooperate. Consider instruction. A parent can instruct a child by doing the task where the child can see the actions; a child who has sufficient mimicking ability will learn simply by repetition. This kind of instruction does not require the parent to be concerned with the child's mental state. But suppose the child does it wrong. Will the parent be able to guide the child's actions? Can the parent decompose the complex action sequence into simpler steps and train the child piecemeal in the components, then supervise the tying together of the components? In order to do this, the parent would have to know something about the child's mental state: that the child lacked the proper knowledge and that the parent's actions could give the child that knowledge. People can do this; other animals cannot (well, maybe the chimpanzee can—primatologists argue about this point).

Donald argues that true communication requires this ability to represent other people's knowledge. Although chimps can mimic behavior, they cannot do the next step: mime. Mime differs from mimicry in that it is intended to transmit information to another. A mimic simply repeats another's actions. A mime performs an intentional act of communication through actions.

When it is time for my wife and me to go out for our daily run, I sometimes communicate this to her through mime. Suppose I am inside the house and my wife outside. I might go to a window, knock on the glass to attract my wife's attention, then point at my watch and run in place for a few seconds. To my wife, the knock indicates a request for attention, the pointing at the watch indicates a concern with time. Couple this with the miming of running and the message is clear: Time to go running.

Miming allows for reasonably complex intentions and states to be conveyed to another, even in the absence of language. This is why Donald postulates the stage of miming as the next level above primate intelligence: It is not as advanced a stage as is possible once language exists, but it is more advanced than even the most advanced animals are capable of.

Miming is a powerful communicative tool, but it is limited in power when compared with language. Mime is limited in the range of ideas that it can convey. Future, hypothetical, and imaginary states are difficult or impossible to represent. Yes, modern mime can transcend these limitations, but this is mime performed by a language-based, educated society, in which established conventions exist.

Language is a powerful representation of ideas. A sentence like "The sky is blue" would seem straightforward, easy to understand, and no different than could be represented by a mime who pointed first at the sky, then at blue-colored objects. But language allows us to go beyond simple statements about world states. We can say, "I hope it doesn't get as hot today as yesterday," thereby setting up a comparison of the not-yet-happened world state of today with the remembered one of yesterday, communicating our hopes and desires. There is no way to express such thoughts without language: Humans can do it; no other animal even comes close.

Complex planning is another mental activity at which humans excel. To plan is to consider several alternative courses of action, weigh the implications of each of those alternatives, compare, then select. Although animals can form simple plans, such as to stack objects on top of one another to form a platform that will enable them to reach objects hanging in the air, chess is beyond their abilities. Identifying just where in the range of intellectual challenges from stacking objects to playing chess their abilities fail is not important: The important thing is that their brain structures do not support such complexity.

Such planning is used in everyday action, in reasoning, and even in choosing how to say some things. As we saw, the study of deceit has revealed much about the limitations of brain power of various animals: Successful deceit at the human level requires planning and analysis. I am happy to say that a powerful intelligence can be used for cooperative, supportive behavior too, not just for deceit.

Donald argues that the analyses required for complex planning, which involve considering and comparing several hypothetical courses of action, are not possible for animals that have progressed only to the mimetic stage of evolution. In my terminology, deep reflective thought is not possible. For these abilities, Donald postulates another evolutionary stage, the mythic stage—the stage of telling connected, coherent stories.

We do not need to prolong this speculation about the exact nature of the evolutionary process nor about the exact comparisons of apes and humans. The important point is that humans have a rich, though limited, representational capacity, one that can be expanded upon through external structures. Mind you, the human brain is also limited in how much advance planning it can do. Our working memory is limited in capacity. It is very difficult to do complex problem solving in our heads. We invented poetry and music, art and science, mathematics and logic. All of these are enhanced through artificial devices, through artifacts. We humans have managed to overcome the limitations of brainpower by inventing external devices to aid in thought. We expand the mind's representational power through the use of external structures and representations, through cognitive artifacts. That is why Donald and I both believe that the real power of the human mind, today and in the future, lies with our technologies. Through technology, we develop external representations and systems that join with our cognitive abilities to provide skills far beyond what can be accomplished through the unaided mind.

HUMAN COGNITION

We are excellent perceptual creatures. Experiential mode is our preferred way of working: See a pattern, immediately understand it. This is what makes the expert so rapid at comprehending, so smooth at responding, so fast at diagnosis. Another common phrase used in psychology to describe this state is "going beyond the information given." A simple fragment of information and we immediately recognize the whole.

How many times have you recognized telephone callers from their very first words? We can often recognize a musical piece from the first few seconds, sometimes even identifying the singer and the band, or the orchestra and the conductor. Sometimes we can identify a friend or relative from a cough or a footstep. This is a marvelous facility, but it can be a dangerous one, for there really is not enough information in that fragment to make a precise identification. The rapid identification process usually works because the number of different events or different people that we are apt to encounter is limited. Moreover, the events are not evenly distributed: Some are a lot more frequent than others. It is those frequent events that we can identify so rapidly. Infrequent events are much more difficult.

The problem with infrequent events is that they are indeed infrequent. When they occur, our fast-working recognizing apparatus is apt to have already sprung into action and classified it—as something we already knew, something that was more frequently encountered. Worse, we are usually so confident that our initial judgment is correct that we seldom even question it. This tendency not only contributes to some types of error, but it then makes it difficult to *discover* error. I return to this point shortly in the discussion of tunnel vision.

Much of our decision making and problem solving is done by analogy, by comparing the current situation with some earlier experience. What kind of earlier experience? One that matches on the major features, one that is available in memory. But human memory is flawed: The things available in memory are apt to have one of two characteristics—they happened recently, or they had some unique, emotional impact. Thus we remember recent events, instances of good fortune, and major calamities. The everyday events, when everything goes normally and nothing of note occurs, are not so easily remembered. Reflect upon that for a moment. If we make decisions based upon our memories, and our memories mainly make available the unique and the recent, what does that mean about the quality of decision making?

The Power of Stories

I have sometimes been amused during my attendance at high-level business meetings in American industry, amused at the discrepancy between the way that we are told that important decisions get made and the truth.

We are told that important decisions should be made logically, with full attention to all the relevant facts, without attention to irrelevancies. Just the facts, please. If only we know the facts, then we can think logically about the possible outcomes and decide intelligently.

In my experience, that's not always how it happens. Oh, yes, important decisions are made surrounded by facts. Each decision maker is provided with a folder filled with material: numbers, spreadsheets, computations, cost estimates. Usually, the decision making is preceded by formal presentations: Eager young executives stand up in front of the board with flip chart, slide projector, and computer. Fancy graphics fill the screens, numbers and formal predictions abound. All is spelled out just as the textbooks say it should be.

I remember just such a meeting of senior executives at a major American company, decision makers inundated with statistics and facts. And then one of the highly placed decision makers spoke up: "You know, my daughter came home the other day and said . . . ," and the story poured out. "Hmm," another executive said, "that makes sense, but you know, the other day . . . " and out came a story. A few stories, some discussion of the stories, and the decision. All those facts and statistics, and in the end, the decision was based upon some personal stories.

There is something important and compelling about stories that bears considering in greater detail. Stories are marvelous means of summarizing experiences, of capturing an event and the surrounding context that seems essential. Stories are important cognitive events, for they encapsulate, into one compact package, information, knowledge, context, and emotion.

The context makes a huge difference. It's as if people make up imaginary scenarios for each event and determine plausible actions, explanations, or responses based upon how well they fit the scenario. "People like to tell stories," says Roger Schank, a cognitive scientist who has tried to understand how memory is used in everyday settings.* The stories we tell not only explain things to others, they explain them to ourselves.

Scientists who study human thought tend to like logic. Indeed, the mathematics of logical thinking was derived, in part, as a prescription for clear and proper thinking: Logic is what people ought to follow when they think, at least so goes the firmly held belief in industrialized society.

We might ask why. After all, logic is artificial: It was invented some two thousand years ago. If logic were a natural property of human thought, people would not have so much difficulty understanding or using it. Logic, however, is a branch of mathematics, not of natural thought. It describes a particular framework for understanding the causal relationships among concepts and events. Logic assumes a certainty to the world: This event is true or false, this concept is (or is not) a part of that concept. This idea implies (or does not imply) that one. Logic provides a powerful reasoning tool. Start with a set of clear assumptions; add a set of clear, precise rules; and the result follows—no ands, ifs, buts, or maybes.

Ah, if only the world were like logic, clear and precise, with no sloppiness or indeterminate states. But it isn't. Moreover, logic is an abstraction. Logical analysis carefully represents only those aspects of the situation that are thought to be important. The decision is based upon those. Stories are suspect because, well, each story presents only a single example, and moreover, the details can overwhelm the telling even if the details are not the most relevant part for the purpose.

The problem is that in its attempt to abstract the relevant from the irrelevant, logic oversimplifies to the extreme. Logical analysis only applies to information that can be measured, but what can be measured and what is important are not necessarily related.* One can measure size and weight, cost and time, but one cannot measure value or beauty, pleasure or pain, moral good or evil. These are all subjective concepts, and even though all may agree that morality, beauty, and pleasure are important, there is no way to translate them into the language of logic, no way without badly distorting their content.*

Stories have the felicitous capacity of capturing exactly those elements that formal decision methods leave out. Logic tries to generalize, to strip the decision making from the specific context, to remove it from subjective emotions. Stories capture the context, capture the emotions. Logic generalizes, stories particularize. Logic allows one to form a detached, global judgment; storytelling allows one to take the personal point of view, to understand the particular impact the decision is apt to have on the people who will be affected by it.

Stories aren't better than logic; logic isn't better than stories. They are distinct; they both emphasize different criteria. I think it very appropriate that both be used in decision-making settings. In fact, I rather like the ordering that often happens, usually accidentally: First the data and the logical analysis, then the stories. Yes, let the personal, emotional side of decision making have the last word.

Error

"To err is human," goes the old folk saying, and the saying is perfectly correct. We err for many reasons, in many circumstances. The reasons for error are starting to be fairly well known, but among them is the simple fact that much of human error is human-imposed.

There are two major classes of error: slips and mistakes. A slip occurs when the action that is performed is not the one intended, such as when someone pours salt into a cup of coffee or perhaps empties the newly poured cup of coffee into the trash and starts to drink the coffee grounds. A mistake occurs when the action that is intended is wrong. Mistakes tend to be much more serious than slips.

People err, that is a fact of life. Another fact is that some situations seem as if they were designed to cause errors, especially when their design fails to take human abilities into account. Many situations seem designed as if to deliberately form a mismatch with human capabilities. Here are some of those capabilities:

- Human memory is well tuned to remember the substance and meaning of events, not the details.
- Humans can essentially attend to only one conscious task at a time. We cannot maintain attention on a task for extended periods. Basically, we are sensitive to changes in the environment: We attend to changing events, not continual, ongoing ones. The same is true for memory: We tend to remember novel and unexpected events better than regular, recurring ones.
- Humans are pattern-recognition animals, matching things that appear similar to past events.

Tunnel Vision

At ten o'clock one morning, I visited my physician to get a prescription for a recurring bout of acute asthma. He prescribed the standard cortisone treatment for such cases: a high immediate dosage followed by a week of rapidly reduced dosages. (The immediate high dosage gets me to a normal state, and then the reductions get me off the drug as soon as possible.) In addition, I was given an antibiotic, since we both felt that the asthma had probably been triggered by a chest or nasal infection: I had some signs of infection plus a history of nasal infections.

Two weeks prior to this episode the same thing had happened: I visited my physician in the morning, and he gave me the same prescription. However, the treatment had not helped, and at 3:00 AM on the day following my office visit, I rushed to the emergency clinic with a high temperature, violent chills, and nausea. After several painful hours and multiple tests, I was released. I was better the next day.

On this second episode, we both assumed that whatever had triggered the earlier episode had not been killed by the treatment. Hence the repetition of a prescription for the previous antibiotic. I left my physician, purchased my medication, and took the first dosage. The medication worked. I went to work and by noon was able to conduct the department faculty meeting. But around 2:00 PM, my temperature started rising. At 4:00 PM, I returned to my physician, who immediately put me in a hospital bed with intravenous tubes and multiple blood tests. I was suffering from a rapidly increasing temperature and violent shaking. My temperature still kept rising, and in the middle of the night, I had to be placed on a refrigerated blanket in order to force my temperature down. Needless to say, I was being heavily medicated.

The strange thing was that despite a battery of sophisticated tests, no signs of infection could be found. I was unable to bring up any more infected sputum (I could barely cough up anything), and my white-blood-cell count was only slightly elevated—not enough to be considered abnormal. None of the various samples extracted from me gave clear signs of problems. No signs of anything were detected in chest and sinus X rays.

I was admitted Wednesday. By Friday morning, I was sufficiently recovered that I was sent home, with yet another tapering dosage of cortisone and another antibiotic (a different one). The diagnosis? Unknown. Well, the resident physician still thought it was pneumonia, with a delayed reaction in order to explain the lack of signs in the lungs. The official diagnosis was "an acute bacterial or viral respiratory infection." In English, this means, "something or other involving the lungs."

On Friday afternoon, a few hours after I was released from the hospital, I felt much better. I hadn't been home since the time I had been admitted to the hospital, so the bathroom countertop was littered with medication—all the new items I had just gotten plus the old pill bottles I hadn't had a chance to put away. I tried to organize my things and take my newly prescribed medicine. Alas, in the process, I inadvertently took the wrong antibiotic—the one that had been prescribed the two earlier times, Bactrim-DS—not the one I was now prescribed, E-Mycin 333. I went to bed, and whoom—at 3:00 AM, I was back in the emergency room with an extremely high temperature and all the rest.

Except this time my error in medication allowed me to figure out the root of the problem. Just before I left for the emergency room, I tried to piece together the story. The wrong antibiotic was the important clue. The previous two times I had also taken the same antibiotic. Now all the evidence fit. I looked up the drugs I had taken in my home copy of the *Physician's Desk Reference:* Yup, my symptoms matched the known side effects on those first two prescriptions of that drug. I was allergic to sulfa-based drugs. Each of the previous times, once I got to the hospital, I was given different antibiotics, so my symptoms disappeared. My symptoms were severe enough that I had to return again to the emergency ward, but this time I was able to explain the reason for my condition to the attending physician (and wonder of all wonders, convince him).

Why did it take so long to diagnose my problem? How did so many people mess up? Why did I myself not correctly identify the culprit?

This phenomenon of failure to correct a misdiagnosis is well known in psychology. I had even studied and written about it, but this was the first time I had experienced it so dramatically for myself. I once gave the phenomenon the rather unwieldy name of *cognitive hysteresis.* This was derived from the phenomenon in magnetism called *hysteresis,* which refers to the property that once a material has been magnetized in one direction, it is difficult to make it change to being magnetized in the opposite direction. Simple magnetic fields that earlier would have caused the unmagnetized substance to be magnetized in either direction no longer have any effect. Once the magnetic state has been set, it is stable and resists further change. This is a very useful phenomenon for the magnetic-recording industry. It makes magnetic recordings possible (by ensuring that the recording will be stable and not easily changed). In earlier generations of computers, the phenomenon was exploited for computer memories: Tiny magnetic cores were magnetized in one direction to represent a value of "1," in the other direction to represent a value of "0." The hysteresis property kept the memories stable even as they were continually buffeted by small magnetic fields. The magnetic properties of hysteresis seem quite analogous to the phenomenon observed with human misdiagnosis.

There are other names for this phenomenon in the psychological research community: *functional fixedness, cognitive narrowing, tunnel vision.* Whatever the name, they all describe pretty much the same phenomenon: People tend to focus upon the active hypothesis and, once focused, find it very difficult to change, even in the face of contradictory evidence.

In industrial accidents, in such industries as nuclear power or aviation or railroad and ocean shipping, the real problem is not the simple human error of reading the wrong instrument or flipping the wrong switch. The real culprit is misdiagnosis. Why? Wrong switches or instrument readings are simple errors, usually with little consequence and usually caught: The wrong deed was done, and almost anyone can spot that fact. But a misdiagnosis is something different. A diagnosis is an interpretation and explanation of events. Experts usually give intelligent diagnoses, even when they are wrong. The diagnosis explains the facts and is judged to be plausible. Misdiagnoses, therefore, are almost always intelligent explanations of what is being experienced. They explain the facts. Thus, as new information comes in, it is interpreted into the picture given by that initial diagnosis. Discrepant information is explained away. The more people involved, the more likely the false trail will be maintained. Each person cleverly thinks of new ways to keep the old hypothesis going.

None of this is deliberate. In fact, overall, it is a very productive way to proceed. As I have already pointed out, one of the strengths of human intelligence is that we can diagnose something rapidly, even before all the evidence is in, then quickly sift the relevant information from the irrelevant. It is this ability that makes us so superior to machines or, for that matter, makes the expert so superior to the novice. The problem is that whenever this wonderful property gets sidetracked by the wrong explanation, it becomes very difficult to get it redirected.

Let me tell you another story, this one from aviation. This story comes from the pilot of a cargo plane who voluntarily submitted the report to NASA's Aviation Safety Reporting System. The pilot was quite experienced; the incident happened while he was flying alone and landing the airplane. Taking off and landing are the periods of highest work load for pilots, especially when there is nobody else in the cockpit to share the work. There are many things to be done, many radio settings and navigation aids to be set, and the visual information out the window is less clear than you might think. They are also the most dangerous phases of flying. The airport usually provides a number of different aids to help pilots identify the runways. All runways are numbered with compass directions. The runway in this incident, runway 35, has a compass direction of 350° and runs almost due north (the same runway, when viewed from the other direction, would be called 17, indicating a compass direction of 170°, almost due south). In addition, each runway has a special light, the visual approach slope indicator, or VASI—two light bars located on the side of the runway about fifty yards apart along the touchdown zone that help identify the runway and guide the pilot to the correct glide slope and orientation. Here are some quotations from the report:

> . . . I sighted the airport and was cleared for a visual approach to Runway 35. . . . I had difficulty lining up with Runway 35 because the lighting seemed poor, but having landed on Runway 35 the three previous times I flew the run, I was confident

> I had the runway. On short final, I noticed there was no edge lighting or VASI, and that the centerline lights were green. Although none of this seemed correct, I was sure I had the runway. I rationalized that the edge lights and VASI had been knocked out by lightning. Airport maintenance vehicles on tower frequency discussing lighting problems lent credence to my theory. It never occurred to me to check the directional gyro. Until the nose-wheel lowered and I saw the yellow painted centerline, I refused to let go of my belief that I was looking at Runway 35. . . . I had landed on taxiway Echo.
>
> I believe the primary causes of this event were my stressful flight around thunderstorms without radar and my mind set. . . . In spite of the evidence—no edge lights, no VASI, and green centerline lights—I refused to believe that it was not Runway 35.

You may think this incredible, but this type of behavior happens frequently. In the nuclear accident at Three Mile Island, it took hours for one misdiagnosis to be discovered, and then only when the work shift changed and a new crew of workers arrived.

This same kind of tunnel vision happened to me: Because everyone was focused on finding the cause of the infection, the possibility that it was not an infection simply never entered anyone's mind—not mine either, until that third incident. I didn't know I was allergic to sulfa drugs—physicians had asked me all my life, and I had always responded, "No, not allergic." It seemed logical that I had an infection. I did have a history of sinus infections, and my asthma must have been triggered by something. On my first visit, I reported yellow nasal drip. On the second, I had a yellow sputum and even showed a sample to my examining physician in the morning. All this suggested an infection. By the time I was in the hospital, everyone wanted a specimen for analysis, but by then, I could no longer cough up anything at all.

Everyone found new explanations for what was happening to me. Three different physicians visited and pondered the charts. Negative test results are not that uncommon. Worse, no test is really that decisive: They come back with all sorts of numbers that can be interpreted. "Hmm," the resident who watched over me in the lonely hours of the morning would say, "your white blood count is slightly elevated. Normally, we would not be concerned with this amount, but today, yes, this is a sign. And," he would continue, remembering some exotica from his medical classes, "I still think you have a form of pneumonia. Not all symptoms of pneumonia show up immediately—I bet we'll see it in the lungs tomorrow."

The conclusion to the story of my own misdiagnosis illustrates the perils of human decision making. Several weeks after my repeated episodes of emergency treatment, I returned to my physician for a general checkup and review of my condition. During the discussion, he said, "You know, it's a good thing you told us about your reaction to that sulfa drug. The other day we had a patient on the ward with those same symptoms and I said, 'Hey, let's look at possible drug reactions.' That's what it was. Same drug even. So you see, your experience helped someone else."

One way of trying to avoid this trap is to have an outside person come in and review things, deliberately questioning every assumption and decision. In part, this is the role now assigned one of the crew in commercial aviation when there is a problem with the airplane. The role is difficult to perform, however, for two reasons. First, it is natural for the new person to fall into the same trap as the others did. Second, unless great care and skill are exercised, a person who questions every statement and action is apt to be judged a troublemaker. It is not an attitude well suited for friendly, cooperative interaction. Here is one case where an intelligent machine has some benefits: It can question the actions and offer advice in a nonthreatening manner. Even if the questions seem unfriendly or ignorant, that is OK, for that is the way of machines. The crew are apt to think the machine an intrusive pain, but every so often, it might turn out to be a useful one.

What we need is a machine that could maintain a rich, interactive database of past statistics and incidents that would be automatically available when making decisions. This would remind us of alternative possibilities and also let us determine the likelihood that those other interpretations might apply. We want to be on the lookout for exceptional circumstances, but we do not want to diagnose every situation as some rare, exotic case.

Why do we make errors? In part, because the machine-centered tasks imposed upon us through our technology ask us to behave in ways incompatible with our fundamental capabilities. Want to guarantee error? Then devise tasks that require using the memory for details, that require devoting extended periods of attention to an unchanging situation. If the environment consists of rows of similar-appearing controls or displays, error in reading or operating them is almost guaranteed to happen.

If the task does not make sense, errors are likely—humans are good at making sense, not at dealing with nonsense. We function by creating mental models—mental explanations of the things we interact with—and if the technology does not provide the information required to create a proper model, we may very well create an improper one. In addition, feedback is essential to keep us informed and focused, and if our tendency to make up explanations is to be kept in synchrony with the actual situation. Highly accurate repetition is not our strength: Imaginative, insightful interpretation is.

People do err, especially when required to do things for which we are not suited. The trick in designing technology is to provide situations that minimize error, that minimize the impact of error, and that maximize the chance of discovering error once it has been committed. The human-centered way.

SIX

DISTRIBUTED COGNITION

MODERN COMMERCIAL AIRPLANES FLY WITH TWO OR THREE PEOPLE IN THE cockpit. One, who sits in the front left-hand seat, is the captain, the person in charge. A second pilot, the first officer, sits in the front right-hand seat. In older aircraft, a third person, the second officer or flight engineer, sits sideways just behind the first officer, facing a panel of controls and displays on the wall of the cockpit. The captain and first officer usually alternate jobs, one flying the airplane during one leg of the trip, the other flying during the next leg, so they also designate themselves by the labels "pilot flying" and "pilot not flying."

The two pilots sit in front of a large panel, the captain's side largely duplicating the first officer's side, with a large control wheel—something like the steering wheel of a car—in front of each pilot. The two wheels are linked, so that whenever one pilot turns one wheel or moves it forward or back, the other wheel follows along. In between the pilots is another set of instruments for controlling the engines, radios, and flaps. These instruments and controls are used by both pilots, so there is only one set.

Control rooms—whether the cockpit of a commercial airliner or an industrial plant—tend to contain great big controls. In power plants, there are huge electrical switches, huge meters that display the state of the plant. Because there may be thousands of controls and displays—in one nuclear power plant that I

Figure 6.1. Cockpit of the Boeing 747-400 Airplane. This is a modern "glass" cockpit, with most of the mechanical gauges of older aircraft replaced by computer-controlled displays. The captain sits in the left chair, the first officer in the right. The control wheels (just in front of each pilot's chair) are yoked—connected so that both move together. Most of the instruments and controls in front of each pilot are duplicated for the other. Many of the instruments and controls in the center are shared. (Photograph courtesy of Boeing Commercial Airplane Group.)

studied, there were an estimated four thousand controls and displays—the rooms are huge, as large as a small house. Several people will normally be monitoring the controls, depending upon the plant and the activity taking place. Large controls in spacious control rooms are the norm in industry. I have seen similar displays and controls in large ships, chemical processing plants, manufacturing plants, and even the control room for one of the lines of the Paris Metro.

The first thought that strikes the modern scientist looking at the controls is that they seem quaint and old-fashioned. When I first saw a nuclear power control room, I was also struck by the thought: "Why on earth does it have to be so big?" Sure, once upon a time, you probably needed a big wheel to turn the rudder of a ship or to operate the control surfaces of an airplane. Once you needed big electrical switches to control all the current that passed through them. You needed big meters and indicators so that they could be seen as the operators walked up and down in the control room. But today none of this is necessary. Most modern equipment is controlled remotely. It is no longer necessary for the control wheel to actually turn the rudder or operate the airplane's wing surfaces. The large lever that controls the landing gear of an airplane no longer actually moves the gear up and down. No, the controls simply send signals to electric or hydraulic motors that do the actual movement.

It would be entirely possible to take the huge room filled with controls for the power plant or the large control panels of the ship and commercial airplane and put them on a small computer: Show the displays on a couple of colorful computer screens and operate the controls with a simple keyboard, a small switch panel, and the ability to turn things on and off just by touching appropriate areas of the screen. Not only could one do this, but it has been done: Excellent examples of these displays and controls exist in the research and development laboratories for all these industries and, for that matter, in the game world, where one can often purchase excellent simulations of the real devices as games for the home computer—simulations that are good enough to be the model for a real control.

The new technologies seemingly eliminate the need for the large controls required by the old-fashioned mechanical technology. The lesson has not been lost on designers. The new airplanes from Airbus have no control wheels. Instead, the two pilots each have small joysticks, not unlike the ones used for computer games. The captain has a small joystick on the left side of the airplane, controlled with the left hand. The first officer has a small joystick on the right side of the airplane, controlled with the right hand. Unlike the control wheels of traditional aircraft, which are interconnected so that one turns along with the other, these two joysticks are independent. They could both be used at once, without either pilot noticing. The airplane's computer decides which one to follow.

Taking this idea a step further, the American National Aeronautics and Space Administration (NASA) has a prototype advanced cockpit in its simulator facilities at the Ames Research Center in California that has a typewriter keyboard in front of each pilot. Make those controls smaller and you could free up a lot of space for the pilots. Then you could even enlarge the windows, so they could see better out the windows as they were flying.

It turns out, though, that those big outdated rooms, those large outdated controls, offer many benefits. The benefits are important to the distributed nature of the job. Although many modern plants and most airplanes can be controlled by a single person, when problems arise, it is valuable to have several people around, the better to share the work load, the better to make decisions.

The critical thing about doing shared tasks is to keep everyone informed about the complete state of things. The technical term for this is *situation awareness*: Each pilot or member of the control team must be fully aware of the situation, of what has happened, what is planned. And here is where those big controls come in handy.

When the captain reaches across the cockpit over to the first officer's side and lowers the landing-gear lever, the motion is obvious: The first officer can see it even without paying conscious attention. The motion not only controls the landing gear, but just as important, it acts as a natural communication between the two pilots, letting both know the action has been done. In fact, the motion helps the captain remember that the task was done: Flip a bunch of tiny switches and it might be hard to remember whether the landing-gear switch was flipped. Lean over and pull down a huge lever and the memory of that muscle movement is distinct and retained. The same with the control wheels: When one pilot moves the controls, the other pilot knows it. Automatically, naturally, without any need for talking.*

Now consider the two small joysticks used in the all-electronic Airbus aircraft. Many who study aviation are very concerned about the unintended side effects of these sticks: The natural communication between the two pilots is lost. There is no way for one pilot to tell whether the other pilot is controlling the airplane except by asking. There have already been instances of confusion, in which each pilot thought the other was controlling the plane, whereas in fact neither was. In other cases, both thought they were in control at the same time. Neither situation is good. The same problems do occur with the control wheels, but those problems can be detected rapidly, for the movement (or lack of movement) of the wheels presents large visible cues. Moreover, it is easy to look over at the other pilot and check for a hand on the wheel or other large controls; it is not so easy to see whether the hand is on the small, side-mounted joystick.

The need for communication and synchronization of actions among members of a team is a very subtle phenomenon. The large mechanical controls and the resulting large control rooms required people to move around a lot as they did their tasks. As a result, a lot of communication was shared, but invisibly, accidentally, without people really being aware that it was happening. Nobody realized just how important this was to the smooth operation of the system until it went away.

A similar situation was observed when my colleague Edwin Hutchins studied the navigation procedures used in large ships in the United States Navy.* Members of the navigation team communicated with one another through telephone handsets, so each could hear what the others were saying. The person taking bearings of landmarks from the port side of the ship could hear the person taking bearings off the starboard side. The chief and the plotters heard everything. Periodically there were errors. The bearer takers were instructed to look for inappropriate landmarks, or the readings were reported or recorded wrong. When equipment broke down, manual corrections had to be applied to the readings given by the magnetic compasses, and during the initial stages of the breakdown, when everyone was under some stress and time pressure, more errors were made.

The normal response of the cognitive scientist to the babble of voices over the telephone sets and the prevalence of error is to try to simplify things, to get rid of the error. Maybe the telephone lines should be connected individually to each member of the team so they wouldn't have to listen to all that irrelevant stuff from the other people. The error rate certainly ought to be worked on: Error can't be a good thing. Wrong.

Hutchins showed that the shared communication channel and, especially, the shared hearing of the errors was critical to the robustness and reliability of the task. A navigation team is a permanent fixture of a ship, but the individual members of the team are continually changing. At any one time, the team is composed of individuals who vary in skill from novices to accomplished experts. The shared communication keeps them all informed. The shared listening to the errors and the corrections acts as an informal, but essential, training program, one that is operating continually and naturally, without disrupting the flow of activity. In fact, two different kinds of people are being trained simultaneously. It is obvious that the person who made the error is being trained. It is not so obvious that the rest of the crew is also learning from the event: The less experienced crew members learn by hearing of the error and listening to the correction activities; the more experienced crew members are learning how to train, noting what kinds of error correction and feedback are effective, what kinds are disruptive. Over the years, as the shipmates change which part of the task they are responsible for, as some members leave and new ones join, this shared communication channel, with its shared teaching and correcting process, keeps everyone at a uniformly high level of expertise.

The unplanned properties of the large control rooms that enhance social communication and the training roles played by the detection and correction of errors teach several lessons. The most important deal with the nature of shared work, shared communication. These are subtle activities, and we still know remarkably little about how this process takes place, about what factors make shared work a pleasant, effective interaction and what factors make it stressful, inefficient, and ineffective.

Many of the essential properties of effective shared action seem to result from "accidental" side effects of the old-fashioned way of doing things. I put the word *accidental* in quotes because I suspect the procedure is not quite as accidental as it might seem, even if it was never consciously designed. That is, over years of experience, the procedures for performing these tasks have gone through a process of natural evolution from their original form to their current shape. Over time, a long sequence of minor changes would occur, each modifying procedures in a small way. Changes that were effective would be apt to stay; changes that were detrimental would be apt to die away. This is a process of natural evolution, and it can lead to remarkably efficient results, even if nobody is in charge, even if nobody is aware of the process.

It is dangerous to make rapid changes in long-existing procedures, no matter how inefficient they may seem at first glance. New technologies can clearly provide improvements over old methods. The old-fashioned control rooms are indeed old-fashioned. Many of their properties, even the ones people grow fond of, are accidental by-products of the technology and may even be detrimental to the task. New technologies can indeed make life more enjoyable and productive. The problem is, it isn't always obvious just which parts are critical to the social, distributed nature of the task, which are irrelevant or detrimental. Until we understand these aspects better, it is best to be cautious.

Natural, smooth, efficient interaction should be the goal of all work situations. Alas, natural interaction is often invisible, unnoticed interaction: We don't know it is there until we remove it, and then it may be too late. We do know that communication is important, however. Listening to the chatter of air traffic controllers turns out to keep pilots informed about all the other airplanes along the route: Replacing this chatter with computer messages sent only to the relevant airplanes destroys this critical aspect of situation awareness, even while giving the benefit of more accurate, less confusing messages. In a similar fashion, replacing the office clerk who delivers mail from department to department with a computer-controlled robot also destroys one channel of communication among departments. Automating factory control or forms processing can also hinder the informal communication processes among workers that allow productive, unofficial decision paths to develop within a company.*

The human side of work activities is what keeps many organizations running smoothly, patching over the continual glitches and faults of the system. Alas, those inevitable glitches and faults are usually undocumented, unknown. As a result, the importance of the human informal communication channels is either unknown, unappreciated, or sometimes even derided as an inefficient and obstructive, non–job-related activity.

Eventually, the natural process of evolution will work even upon the latest of technologies. The problem is that in the meantime, if we are too precipitous in making change solely because it is possible, we are apt to run into difficulties. When these difficulties occur in commercial aircraft or large industrial plants, the results can be tragic.

DISEMBODIED INTELLIGENCE

The sciences of cognition have tended to examine a disembodied intelligence, a pure intelligence isolated from the world. It is time to question this approach, to provide a critique of pure reason, if you will. Humans operate within the physical world. We use the physical world and one another as sources of information, as reminders, and in general as extensions of our own knowledge and reasoning systems. People operate as a type of distributed intelligence, where much of our intelligent behavior results from the interaction of mental processes with the objects and constraints of the world and where much behavior takes place through a cooperative process with others.

In the research areas studied by experimental psychologists, linguists, and workers in the field of artificial intelligence, thought and understanding are assumed to take place with little or no hesitation, little or no error, and little or no doubt. Scientists make these assumptions in order to simplify their task. "After all," they will state, "the phenomena we are studying are so complex that it is essential to look at them first without all those other complicating factors. Then, after we have understood the isolated case, we can move on to the more realistic and complex situations." The problem with this point of view is that the so-called simplification may be making the task more difficult.

With a disembodied intellect, isolated from the world, intelligent behavior requires a tremendous amount of knowledge, lots of deep planning and decision making, and efficient memory storage and retrieval. When the intellect is tightly coupled to the world, decision making and action can take place within the context established by the physical environment, where the structures can often act as a distributed intelligence, taking some of the memory and computational burden off the human. To give one example: Linguists are continually worried about the amount of ambiguity that exists within language. A huge amount of scientific research has gone into developing schemes for understanding and trying to minimize this ambiguity. But the ambiguity almost always results from the analysis of single, isolated sentences: in real situations, where several interacting people deal with real events, the sentences usually have only one meaningful interpretation. Actually, even when communications are ambiguous, they are usually not perceived as such by either speaker or listener, even though both may have different interpretations of the meaning. It is this lack of perception of ambiguity that is important, and it derives from the communicative, social nature of language, something that is entirely missed when the language is studied as isolated, "simplified" printed sentences or utterances, completely abstracted from the real, social setting.

Information in the world can be thought of as a kind of storehouse of data. This has many advantages. The world remembers things for us, just by being there. When we need a particular piece of information, we simply look around, and there it is. Do I need to repair my car? I don't have to remember the exact shape of the part, because when the time comes for me to do the task, the shape is there in front of me. This eases the burden on initial data collection, eases the requirements on learning and memory, and avoids the need for complex indexing or retrieval schemes. Moreover, it guarantees that the values so obtained will be the most timely available at the moment of need.

Of course, it is important to plan ahead, but postponing decisions until the point of action can simplify the thought processes: Many alternatives that would have had to be thought of ahead of time will turn out not to be relevant. Moreover, the physical structures available in the world can then guide the selection of relevant choices.

Approaches to reasoning and planning that rely heavily upon thought, and therefore internal information, run into fundamental problems:

- *Lack of completeness:* In most real tasks, it simply isn't possible to know everything that is relevant.
- *Lack of precision:* There is no way that we can have precise, accurate information about every single relevant variable.
- *Inability to keep up with change:* What holds at one moment may not apply at another. The real world is dynamic, and even if precise, complete information were available at one point in time, by the time the action is required, things will have changed.
- *A heavy memory load:* To know all that is relevant in a complex situation requires large amounts of information. Even if you could imagine learning everything, imagine the difficulty you'd have finding the relevant stuff just when it is needed. Timely access to the information becomes the bottleneck.
- *A heavy computational load:* Even if all the relevant variables were known with adequate precision, the computational burden required to take them all properly into account would be onerous.

The negative side of this is that these world-based decisions must be made and actions must be taken quickly, which can cause oversimplification and incomplete analysis. We all know that actions taken in haste are often wrong actions. With time pressures, there is limited opportunity to consider alternatives or to reflect upon all of the consequences. Clearly, we need to plan ahead, but not to follow those plans rigidly. We need to respond to the situation, to be flexible in the face of unexpected occurrences, to change our activities as the world dictates.*

IN THE WORLD, IMPOSSIBLE THINGS ARE IMPOSSIBLE

The world has an important property: In the real world, it is not possible to do actions that are not possible.* This sounds trivial and obvious, but it has some profound implications when we move into the artificial world of cognitive artifacts. Thus it is certainly not trivial to those who write computer programs that mimic the world. Much of the effort of writing programs that simulate the world must be devoted to ensuring that the simulation cannot do impossible things.

I have flown in extremely sophisticated simulations of aircraft, ones that barely could be distinguished from the real thing. These professional simulators were constructed from real cockpits, they vibrated and sounded like real planes, and moved about two meters in all directions so they could simulate most of the body sensations. And when you looked out the window, you saw the appropriate sights. Yes, the planes behaved just right. But I once flew in a 727 simulator around the streets of San Francisco, a commercial airline pilot at the controls, flying around the Transamerica building. Oops, we flew through the Transamerica building. Not even a tremble. Once we dove into the ground a close to supersonic speeds. Those of us in the cockpit felt somewhat nauseous: Our minds expected sights, sounds, and movements that did not occur. The computer simulator just kept going. Buildings, walls, even the ground are just numerical and graphic abstractions: To a simulator, there is nothing impossible about being 1 meter below the ground.

Suppose a programmer of computer games developed an exploratory game with Harjimé, the protagonist, wandering through the halls of the enchanted castle. Writing the part of the program that controls Harjimé isn't all that difficult, nor is the part of the program that simulates the castle. Want to simulate the room with the hidden treasure? Just draw in the locations of the walls, furniture, secret keys and panels, and the hidden door. But making the simulation work would be a complex task. The hard part is to make sure that Harjimé doesn't walk through the physical objects in the room. If Harjimé picks something up and then puts it down somewhere else, the programmer has to worry about whether there is a supporting structure at the new location, or if the object will fall, tilt, or slide. Harjimé's movements would also have to be carefully monitored to make sure there was always a supporting floor or surface. Harjimé couldn't move up or down unless there was always a suitable support (but he would have to move up, down, or around when he encountered stairs, ramps, furniture, and elevators). Although the task of programming Harjimé and the castle could be given to novices, the task of programming the interaction of the two is complex and difficult enough that it would tax even the experts: How quickly the program could actually execute would be determined by how well it could compute the necessary constraints and interactions.

The point is that in the real world, the natural laws of physics allow only the appropriate things to happen. There is no need to compute whether you are walking through a wall: You simply can't do it. In the artificial world of computer simulation, much of the computational effort goes into the part that results from the artificiality of the situation.

It has long been noted that in dreams, people are free of the constraints of everyday life. We can visualize doing things that are impossible in the real world. Ah, the freedom of dreams, the fantasies released. The impossible actions of dreams might be ways by which people satisfy their fantasies. But they might also result from the impoverished programming power of the human mind.

Suppose, just suppose, that the wonderfully creative fantasies of our dreams are artifacts, accidents of the fact that our minds can't quite handle the computational job of doing accurate simulation. A dream, after all, is a simulation of human action within a simulated environment. The simulation program is executed within the human mind—a disembodied mind, however, for the sense organs are inhibited and the voluntary muscle system inoperative. Consider what it would take to run this simulation properly. The people and objects would have to be created and their actions determined. The environment would have to be created. And finally, the interactions among all the objects and people and the environment would have to be simulated, which means continual checks to ensure that two objects don't pass through one another, that the force of gravity worked properly, and that impossible actions did not take place. It would be a complex programming job, and one that put enormous computational demands upon the brain.

How much easier to simplify the computations. Let objects

pass through one another. Let gravity work in inadequate ways. Free yourself from the constraints of the real world and you reap enormous benefits. The effort is much reduced and the result much more intriguing. Now the human interpretive system can go to work on the products of its own (inadequate?) simulation. It frees up the creative spirits, allows us to contemplate the impossible, amuses and entertains, and creates the industry of dream analysis. Imagine, all these side effects result from simplifying the computational load of the simulation.

WHY ACCURACY IS NOT ALWAYS IMPORTANT

In the days of oral tradition, before reading and writing were widespread, it used to be common for storytellers to go from village to village, telling stories, passing news from one place to another. Here what was important was style and content. These storytellers were famed for prodigious feats of memory, for they could often tell stories that lasted for hours to an enthralled audience. The stories were all memorized. And when modern scholars studied the few remaining storytellers in the few remaining preliterate cultures, they were proudly told how accurate the memory of these storytellers was.

But when the stories were tape-recorded and compared, any particular story varied tremendously from telling to telling. Where was the accuracy? One telling might be twice as long as another. Yet to a villager who had heard both renditions, they were identical, except that perhaps one was better than the other.

To the listener and teller both, word-for-word accuracy was unnecessary. The very notion is not even understood by a completely oral culture. It is only with the advent of writing and tape recorders that we care about such things. It is only the scholar who carefully writes every word of one telling and compares it, word for word, with the next. As for the rest of us, in our normal group settings and activities, who notices, or even cares?

The storytellers didn't memorize the stories, at least not in the sense of the word-for-word learning that we call memorizing today. Basically, the story framework was learned, plus formulas for fill-ing out the phrases and color. The fact that the tales were told in poetry helped, for this put further constraints on the possible wording: The story had to follow the story line, fit the well-known formulas, and fit the meter and rhyme of the poem. The storytellers were able to construct the story anew for each telling, varying it according to the characteristics of the audience. But still, it was the same story, and the listener who heard it once when it lasted one hour would insist it was the same thing as heard the previous week or year when it might have lasted two hours. It was the same story, except for the details of the telling. The fact is, we are social, interacting people, always alert to interpretations, meanings, and reasons. We need stories and context. Who cares whether the details vary? Who cares whether there is word-for-word accuracy? That is simply not important for everyday life.

Human memory is organized around the important things in life: the excitement, the meaning, and the experience itself. Word-for-word accuracy is simply not important, and it is difficult to accomplish. However, this is no longer true in today's technological world. Great accuracy is required. Lawyers watch every step. Machines are sensitive to every deviance. We are forced to use memory in ways not natural to its evolutionary biological history. And so we must turn to artifacts.

Beware: Using artifacts—technology—to help overcome the frailty of human memory may move us in undesired directions and swamp us with excessive amounts of excessively precise information. The question "What can technology do to help?" is almost always the wrong question. Sure, we can devise technological solutions to the problem. Maybe we can invent small, powerful computers that will remember for us, computers small enough to be available at all times. If not computers, tiny voice recorders small enough to be worn on the wrist. But once we start thinking this way we become trapped in an ever-lengthening chain of technology dependence that in turn forces us to deal with an ever-increasing load of detailed information. Because we can't readily grasp all of this, we will need to devise additional technology to aid us, putting us even more at the mercy of our machines. The whole solution is wrong because the problem is wrong. The correct approach is to structure the world so that we do not have to remember such mindless trivia. Then the question of technological aids would never have been asked. No "solution" would have been necessary.

This is the lesson from the preceding sections of the chapter. Those large control rooms may be unnecessary today, but in changing them, we must be sensitive to the social communication that they afford: Changing the equipment may accidentally destroy the informal communication channels that make work proceed smoothly, synchronized among a group of workers without the need for direct verbal communication.

In airplanes and navy ships, shared communication may at first seem unnecessary, exposing people to irrelevant messages. But, the messages carry information about the activities of others, information that at times is essential to the smooth synchronization of the task or, as in the case of the ship navigators, information that serves as an efficient training device for the entire crew, regardless of their level of expertise. People are effective when they work in a rich, varied environment. A disembodied intelligence is deprived of rich sources of information.

Finally, some aspects of technology expose us to demands for accuracy and precision that are of little importance to normal life. Nonetheless, we have altered our lives to give in to the machine-centered focus on high accuracy, even where accuracy is not critical. Our goal should be to develop human-centered activities, to make the environment and the task fit the person, not the other way around.

SEVEN

A PLACE FOR EVERYTHING, AND EVERYTHING IN ITS PLACE

Figure 7.1 The Wooton patent desk. "With this Desk a man absolutely has *no excuse for slovenly habits* in the disposal of his numerous papers, and the man of method may here realize that pleasure and comfort which is only to be attained in the verification of the maxim, '*a place for everything, and everything in its place.*'

"The operator having arranged and classified his books, papers, etc., seats himself for business at the writing table, and realizes at once that he is '*master of the situation.*' Every portion of his desk is accessible without change of position, and all immediately before the eye. Here he discovers that perfect system and order can be attained, confusion avoided, time saved, vexations spared, dispatch in the transaction of business facilitated, and peace of mind promoted to the daily routine of business." Advertisement by the Wooton Desk Manufacturing Company, around 1880. (Italics in the original.) Cited in Showalter and Driesbach (1983). (*Photograph courtesy of the Oakland Museum History Department, Oakland, California.*)

I do not know when people first started complaining about information overload, but at the end of the 1800s, it was a serious problem. Worldwide trade existed. Messages were sent by mail or courier, although they could take weeks or months to arrive at distant locations. The newly introduced telegraph allowed messages to be sent almost instantaneously to distant places, although these were limited in length because the serial transmission of each character by Morse code was slow: It took about a second to send a word, so transmitting a text the size of this printed page might take as long as five minutes. In not too many years, the automobile, the telephone, and the airplane would increase transportation and communication rates. Radio would follow not too far behind. Soon business and commerce could be conducted around the world with rapid, efficient transportation and communication. The resulting changes overwhelmed many companies. They had to figure out how to accommodate themselves to working at a distance. Salespeople had to communicate customer needs to the factory and determine whether the factory could meet them at the desired time and price.

The telephone allowed a remarkable and unexpected change in the conduct of business. Prior to the adoption of the telephone, all parts of a business had to be at the same location: the officers, the manufacturing, the sales, and the shipping all worked together, in the same building. Whenever one group had to communicate with another, they would simply walk over and meet. This kept operations efficient, but small. The sales team and distributors had to be located at a distance, and this created difficulty because of the slow speed of communication by mail and the limited ability to communicate by telegraph. With the telephone, the different parts of the business could be at different locations. The sales and distribution force could be in instant contact. The business could grow in size. This was one of the most profound—and unexpected—impacts of the introduction of the telephone.

The telephone was fine for keeping in touch, but the lack of a written record was a problem. However, the mail systems also improved, in both speed and reliability, especially with the expansion of the railroad system. Now written records could be sent to distant locations. As companies expanded to serve larger geographic areas, they needed better record keeping, especially as the amount of correspondence increased. Copying machines did not exist nor typewriters nor carbon paper. All letters were handwritten, and copies could be made only by rewriting the original.

All sorts of solutions to the problem of making copies were proposed, from pantographs, which duplicated the original writing motion on other paper, to Thomas Edison's proposal of a vibrating pen that could make an impression through several layers of paper. The most successful invention was the "press book," a bound book of blank tissue paper. You placed the original letter under a moistened blank page of tissue paper, then pressed the two together hard with a screw-wheel device. The wet tissue picked up some of the ink from the original, and the resulting mirror-image impression of the text could be read through the back side of the tissue paper.

Press books became a popular way for businesses to keep copies of outgoing letters. This led to the method of recording outgoing mail by date and time—that is, organized by whenever the copies were made. Copies of the outgoing mail were stored in press books in the order in which they were pressed. Incoming mail was stored elsewhere, usually in boxes. Invoices and orders and internal memos either didn't exist or were stored in yet other boxes.

The resulting rapid rise in records and correspondence threatened to swamp offices. Where to put all the correspondence, all the statistics? And how to find the critical information when needed? All sorts of organizational schemes were tried. One problem was that records were bound in ledgers and press books. Because the books were organized by time, with new information being added at the next available place in the book, it was very difficult to find things. Want to track a customer's order? Well, when was the order received? It would have been stored in a box, organized by date of arrival. You had to find the proper box laboriously and search through it. When was the order processed? The paperwork would be stored in yet another box or a ledger: Find it and search. When was a confirmation letter sent out? A copy should be available in the press book for the relevant date. The chore of bringing together all the information became unmanageable.

In their desperate attempt to cope with their storage and retrieval dilemma, the workers began storing papers in boxes, then adding numerous small compartments to desks—pigeonholes—in which the desk owners could try to organize related material. Pigeonhole desks got more and more elaborate until they reached their peak in the "Wooton patent desk," shown as the opening illustration to this chapter: an overwhelming number of slots and drawers of varying sizes. "A place for everything, and everything in its place" went the advertisement. It was an obvious technology, but inappropriate to the task. One wonders how the owners could ever remember where stuff was. The answer appears to be "They couldn't," and the desk is today a museum piece and a collector's item, not the answer to the information explosion.

The multiple slots and drawers of the desk do serve an organizational function: With sixty to one hundred different places, the items to be stored can be divided into sixty to one hundred categories, one category per location. In theory, this keeps any particular item distinct from others, lumped together only with other items to which it is related.

The problem is that so many categories exceed the number of places that a person can easily remember. Three slots, fine. Five slots, maybe. Ten slots, no. With ten slots, perhaps the users will remember what is in the end ones, but they will almost certainly confuse the middle ones. Sixty slots? Forget it. The desk provided little aid to organizational structure, little aid to higher-order categories, and little aid to search. Sure, it was possible to put "pending bills" in one slot and "bank records" in another, but what links these and related categories into a higher-order structure of "financial transactions"? One could try to put related items near one another, but how? Suppose "pending bills" consisted of a large number of small envelopes and "bank records" consisted of several large checkbooks: The physical construction of the desk restricted where these could be put.

Finally, the desk provides no external aids to help users remember those categories. Yes, most of these desks have label holders, but in all the pigeonhole desks I have examined (some in museums, some still being used by their proud owners), I saw very few labels.

Labeling is a nuisance, and the desk provides little aid to this essential task. Even if labels were in place, it would be difficult to scan sixty to one hundred labels. But even finding the correct pigeonhole or drawer does not solve the problem. When the contents of the slot are extracted or the drawer opened, it was difficult even to figure out whether the item was there or not. People tried all sorts of tricks to help, including writing little notes to themselves—indexes—on the front of envelopes so they could tell what was in them without opening them. But the task of writing and maintaining the indexes increased the work load even more.

The pigeonhole desk provides surprisingly little assistance because it puts most of the burden of organizing the material on the user: Too much knowledge has to be retained within the user's head. The result is error and confusion. Consequently, these desks are hard to share with others. If you forget how you organized something, you might as well forget the items. I asked owners of modern pigeonhole desks how they managed. They all liked their desks but admitted that they could only remember approximate locations. And, yes, sometimes they were forced to look in every drawer, every slot until they found the item they were searching for. These desks are intriguing to behold, but they are not effective at their intended purpose.

> A place for everything, and everything in its place
> —if only you can keep track of the places.

THE VERTICAL FILING CABINET: A TECHNOLOGICAL BREAKTHROUGH

It is difficult today to realize that the common, everyday office filing cabinet represented a major revolution in handling information when it was first introduced in the early twentieth century.* It allowed large amounts of material to be organized in an efficient, well-structured manner. Correspondence could be structured by topic or person. The vertical arrangement of the files made it possible to scan them quickly. Various organizational schemes were invented, the simplest and most effective being labeled tabs that divided the files into regions, often organized by function, date, or alphabet. The cabinet allowed individual pieces of paper to be inserted into folders, labeled by topic. This allowed hierarchical labeling: One filing cabinet (or even a group of filing cabinets) might stand for one aspect of the business, a different group for a different aspect. Then, for each cabinet every drawer could be labeled, and within each drawer, the area could be divided into labeled sections, each section containing a set of labeled folders. It was finally possible to organize things systematically, by some agreed-upon procedure.

The vertical filing cabinet wasn't practical until the copying problem was solved. Today we have the xerographic copier, which overwhelms with its ability to create duplicates of everything. Even once the copying problem was improved (by the development first of carbon paper, eventually of copying machines), there still had to be other developments before the filing problem was eased, such as standards for paper sizes (notice how the Wooton patent desk has a wide variety of drawer and slot sizes?). And even once paper sizes were standardized, people had to learn how to file. Courses on methods for filing became a booming industry in the early twentieth century: Without effective filing methods, it would still be impossible to find individual records.

Today, with the existence of many organizational aids, especially computer databases and retrieval systems, it is difficult to appreciate the power of the filing cabinet. But it is a powerful cognitive artifact. Even if not glamorous, it is a reliable, robust technology that really did revolutionize the practice of business.

Just as it was easy to see why the pigeonhole desk fails as an organizer, it is easy to see why the vertical filing cabinet succeeds. Whereas the desk provides a fixed number of locations with no aids to organization, the filing cabinet provides an unlimited number of positions with few constraints on size except that the items all have to fit within the standardized folder. The pigeonhole desk offers no higher-order organizational aids, but the filing cabinet provides many levels of organizational structure. Look at Figure 7.2, a filing cabinet drawer: Colored labels divide the drawer into several large categories, and within each category,

Figure 7.2. The well-organized filing cabinet. Illustrating the wide variety of methods available to organize and label materials in the filing cabinet, not possible with pigeonhole desks.

further labels form subclasses. Then, within each subclass, there can be several file folders, each labeled, each containing relevant material. Even higher order organization is possible. Each drawer can be a different category, as can the cabinets, usually consisting of two through five drawers.

Filing cabinets are not the complete solution, especially in situations where the amount of information that must be stored fills tens, hundreds, or even thousands of cabinets. But for the modest amounts of information accumulated by a single individual or office, they offer a dramatic enhancement over the impoverished facilities of the pigeonhole desk. Modern business would not have been possible without this most important cognitive artifact.

Today we have a variety of organizational methods available to the individual office worker. Each individual can select from a large number of methods to organize records and keep track of activities. While on a trip to Cambridge, Massachusetts, I met a colleague who, when she learned of my interest in organizational methods, described how she organized her office. I was so struck by the efficient profusion of organizational aids that I got permission to photograph the office. So here it is, the modern, well-equipped office, using all the fruits of modern technology, from powerful computer workstation to piles on the floor, wall calendars and noteboards using Post-it Notes and other modern inventions, and even a special cabinet devoted to current, ongoing work, organized as separate piles, one per category—a "pile cabinet."

I asked my colleague (here given the pseudonym of "Chris") to provide me with a written description of her organizational scheme. Here is Chris's reply, sent to me by electronic mail, edited only for clarity and to disguise her identity (all with her permission).

From: chris@xxxx.edu
To: dnorman@ucsd.edu
Subject: My office . . .

You asked me to describe how I organize my office. Here it is:

I've worked pretty hard to make my office environment help me as much as possible.

Much of the office organization is designed to help me find things and remember to do things. I also want to remind myself to work on "important" things, like writing papers, that might easily be pushed aside by less important but more time-critical items. My desk is arranged in an L shape, and I have placed a large calendar on one wall and a large "to-do" list on the other wall. (I also have a special calendar on my door to remind me about the most critical items I have left to do on my major project, which I can read from my desk and see whenever I come in.)

The calendar shows the current month and lists things that I have to do, including meetings, deadlines, and reminders, such as "drop off Chapter 3" or "send E-mail to Executive Committee." I use Post-it Notes, so that I can easily deal with

Figure 7.3 The pile cabinet. Rapid, easy access—as long as there aren't too many piles. The cabinet on top is from Chris's office, the one on the bottom my home office. Both offices also have other organizational tools, including bulletin boards—less structured than the pile cabinet, but offering fast access for a limited number of items.

changes. The calendar is arranged chronologically and includes items that are both important and rather unimportant.

I update the "to-do" list every 3 or 4 months, whatever seems reasonable. This shows all the major activities that I want to accomplish in the next quarter. I think of it as a plan for writing a quarterly report--I want to be able to say that I accomplished everything on the list. (I usually accomplish about 85% and roll over the rest.) The most important items on this list are talks and papers, but I also list reviews, travel, reports, items from home (e.g., income tax), and activities for the professional society that I am now president of.

These items usually require a minimum of 3–4 hours. I use Post-it Notes, which get a large check mark when I'm done with the item. I am careful to leave these up, as a reminder that I'm really getting something done. When I set up my next 3–4-month interval, I take off the items that have been checked and place them in a random collection on the side of the "to-do" list, as a reminder of what I actually got done. This turns out to be very important to me psychologically (if I may indulge in a little self-psychoanalysis)--constantly displaying a list of accomplishments helps me want to accomplish more and helps make the tradeoff of putting smaller items aside in order to add another check mark.

I also have a 3-level filing system for handling papers. My desk is my working surface--I don't mind having it messy while I'm working on a project, but I like to clear it off pretty regularly, probably once every 3–4 days. (Longer if I'm working on a paper.) On the side of the desk with the calendar, I keep items that I use constantly, including the phone, answering machine, tape, stapler, stamps, Rolodex, Post-it Notes in various colors, pens, dictionary, Strunk & White, and thesaurus, and, for aesthetic reasons, a very pretty kaleidoscope. I also have a homemade envelope attached to a file cabinet wall that contains address labels and another for receipts. I use "slash folders" in different colors to keep track of different things (e.g., papers, letters to write, reviews, a stolen car radio claim) and keep the "current" ones on my desk. I also keep empty slash folders behind them.

In my desk, I have 6 drawers: one holds envelopes and blank pads of paper, which I use whenever I want to jot down ideas. Another holds food (tea, soup, spoons, etc.) and medicine. Another holds my finances (checkbooks, mortgage envelopes, calculator, old calendars and checks), emergency greeting cards (!), old letters, small notepads. Another holds basic stationery supplies (paper clips, pens, Wite-out, subway tokens, staples, etc.). Another holds less commonly used items (business cards, file folder labels, file cabinet keys, marker pens, etc.). The last is a file drawer that contains miscellaneous commonly used files (expenses, résumé, current bills and pay stubs, forms, administrative information from my employer). I clear out extra files from here about once a year, usually after I've finished a major project and am looking for something nonintellectual to do.

The other side of my desk, under my "to-do" list, has a notebook full of 35mm slides for talks, blank transparencies, phone books, software manuals, and, for some reason, videodiscs. (I guess because they're too tall to fit on my bookshelf.) I have several stacking "in-boxes," which get used for phone lists, copies, fax info, "political documents," and an out-box, generally for outgoing mail. I don't use an in-box, per se.

I also have a lot of file cabinets. I have a 2-drawer file cabinet on the left of my desk but rarely use the files. (It used to contain my current writing projects, but those have been shifted to the pile cabinet. Now it mostly contains projects that were started but never finished.) I have two more 4-drawer file cabinets to the right of my desk. I use three of the drawers all the time. Two contain a chronological listing of my papers, including extra copies to give to people who visit. One contains current files for the professional society.

I have another file drawer for ancient society files and several others full of old files from work. I look in these drawers about once every 3–4 months.

The pile cabinet has turned out to be a very successful way to deal with current projects. I started using it about 3 years ago. I found that if I put things into file drawers, I'd forget them. Yet my desk isn't big enough to handle everything that I'm currently working on. I converted a cabinet designed to display brochures and labeled it with Post-it Notes. (I generally rethink the categories when I update my "to-do" list.) My current set of projects includes several piles of current society things, such as things to sign, finances, local groups, etc. I also have documentation, specs, meeting notes, etc., from two different software projects, and I have several piles devoted to different aspects of my current project (committee notes, comments from colleagues, the original proposal, different chapters). I also keep different sets of data that I'm working on and have another pile for relevant articles. (This is *hard* to keep to a manageable size.)

I have 5 bookshelves, which hold my books, articles I someday plan to read (!), and journals. I also have moved about 4 shelves full of articles and other items to shelves outside my office.

I have a whiteboard for scribbling ideas and making lists. I also keep a few figures there all the time so that I don't have to redraw them when I talk about them. I also have a flip-chart pad behind the door and periodically use it to think through ideas that I want to keep. I tape these up on the walls, to keep me thinking about the ideas until I have a chance to write them down.

Chris's office illustrates the wide variety of cognitive artifacts available to aid in the organizational issues that dominate modern professional life. She uses several noteboards and chalkboards and several distinctly different calendars. A computer workstation provides some assistance but is also another source of excess information. Post-it Notes, the decade's most important cognitive artifact, play key roles in organizing, flagging, and reminding, after which, as finished items on Post-it Notes are placed in a "completed" area, they serve as a motivational tool to reward the steady pace of activities.

Note the deliberate use of two forms of storage cabinets for files: one for efficient organization, the other for efficient retrieval. One, a traditional "filing cabinet," is used for material that is referred to relatively infrequently. The filing cabinet has the virtue of allowing the use of powerful organizational tools (as you can see from Figure 7.2, the inside of one of Chris's file drawers). The other, a "pile cabinet," consists of horizontal shelves with piles of folders and papers. Pile cabinets lack efficient organizational aids. In the language of organizational structures, file cabinets provide deep, hierarchical representational structures, whereas pile cabinets allow only a shallow, flat structure with just one level of organization: the name of each pile. In this manner, pile cabinets are like the structure of the Wooton patent desk.

This is yet another example of the common tradeoffs inherent in design: Optimization for one feature tends to come at the cost of inefficiency for another. Pile cabinets permit immediate access, but with poor organizational structure: Piles work well for material in frequent use. Filing cabinets provide efficient organizational structure, but getting to a document, even a frequently used one, requires going through several layers of organizational structure.

Despite Chris's attempt at order and structure, several violations are apparent. Some things seem to be placed for historical reasons, not for practical organizational reasons. Why are videodiscs located in an obscure place? "I guess because they're too tall to fit on my bookshelf." Some structures are there for motivational reasons, not just for efficiency. My favorite among these is the Post-it Notes on the calendar for items to be done, which, when finished, are not thrown away (as I would have done) but instead are given a big check mark and moved to a "done" region of the calendar—a constant motivator and reminder that tasks really do get completed.

Chris represents one extreme of organizational efficiency, perhaps more than most people could cope with. But her methods show what a single person can accomplish simply through effective use of cognitive tools.

ORGANIZING KNOWLEDGE

Have you ever wondered why dictionaries and encyclopedias are arranged in alphabetical order? The alphabet provides a rather bizarre ordering: Nearby words are apt to have no relationship to one another. Wouldn't it be better to have a functional ordering, like a department store or a hardware store? Or even like the library? Then nearby items would be related. Doesn't that make more sense?

Can you imagine a hardware store where all the items were arranged in alphabetical order? Ask a clerk for pliers and get sent over to the Ps. But suppose the pliers you wanted wasn't there? You ask a clerk, who says, "Oh, you mean a needle-nosed pliers. That's in the Ns." It just wouldn't work.

A superb example of how well functional organization works is provided by McGuckin's Hardware Store, in Boulder, Colorado, an establishment that carries over three hundred thousand items—more than the number of entries in many dictionaries. Dictionaries seem to think they need alphabetical order, but not McGuckin's. How does the staff manage? How can a customer find the desired item?

Brent Reeves and Gerhard Fischer, in the computer science department at the University of Colorado, wondered the same thing, so they studied McGuckin's Hardware Store, conveniently located near their university.

McGuckin's organization is hierarchical, by function: The class of item being sought determines the section of the store, and within that section, the function determines the subsection. The most interesting cases are when customers start with only vague ideas of what they want. They find a clerk and describe their problem. Clerks know their own area of specialty in exquisite detail but have incomplete or vague knowledge of the rest of the store. That's

OK, as long as the clerk knows enough about the rest of the store to send the customer off in the correct direction. The customer heads off, and after a while finds another clerk, who again redirects the customer. Usually, it only takes one or two encounters before the customer has been directed to the proper section, where section specialists take over. The specialists not only know their small areas intimately, but they are often experts at the tasks associated with the items. Thus, when I visited the store, I came across an item I had searched for in stores in my home city without success (a special fuel bottle for our tiny camping stove). I had always been told that this item no longer existed. When I told this to the area specialist, I was pleased to discover that not only did the clerk know about the item but he could tell me why it was so difficult to find, could describe the competing product that had driven it off the market, and could then tell me where the item was still manufactured in small quantities. Furthermore, he said, McGuckin's intended to continue carrying it: "I like it myself," said the clerk.

This kind of expert knowledge by the specialists allows them to discuss the customer's needs in sufficient detail to transform the initial vague, imprecise description of the desired item into a specific suggested object. Once the object is in hand, the customer and clerk can discuss the actual application and further refine the description, perhaps examining several other objects until the desired one is found. I listened in on one conversation where the customer had come in looking for a fastener. After the clerk and the customer discussed the application for a while, the clerk concluded that glue would be superior and sent the customer off to a glue specialist.

McGuckin's is organized by function: Related items are stored near one another. The people who use McGuckin's do not need to know the structure of the entire store. Once they learn where things of interest are located, they can be assured that related items will be nearby. It is only when customers seek novel items that they need the aid of a clerk. These clerks play the role of what we in cognitive science would call "an intelligent agent": a knowledgeable specialist who can aid in the completion of a task. These "agents" have incomplete or vague knowledge of the rest of the store—just enough to direct the customer toward an agent who is likely to specialize in the customer's needs.

Why are dictionaries organized the way they are? Well, consider trying to find something in McGuckin's without the aid of those agents. The problem for the dictionary or thesaurus or encyclopedia is to figure out how to organize things when there is no intelligent specialist to help the reader. What are the alternatives? When dictionaries and encyclopedias were first invented, they were organized functionally: Related items were put next to one another. The problem was that any single item might be related to many others, but the printed page wouldn't allow all the related topics to be near one another. And what happened when something was related to several items? In the end, the alphabet was the best neutral organizational scheme. If you know the item you are searching for, and if you know how to spell it, then you can use the alphabet rapidly and efficiently. But as all of us must have experienced, the access scheme is flawed.

When we know the word and seek the meaning, then the alphabetical arrangement serves us well. But when we seek alternative words with similar meanings, then a structure that puts related words near one another would be preferable. Sometimes we seek words within a common category (kitchen words or sports terms), in which case a categorical structure would be useful. The same is true with encyclopedias.

Printed books are limited in their organizational structure. As a result, reference books have tended to compromise with alphabetical arrangement of terms. This only works if you know the term under which the item of interest is located, and so in order to aid the user, considerable effort has to be given to providing other entry points: The book's contents are summarized in lists of contents, lists of chapter headings, and sometimes a variety of indexes. Any given part of the text might be "pointed at" from several different places in the indexes and table of contents. Even so, this is often quite unsatisfactory, and the users of reference works often end up keeping many places in view at the same time, usually trying to do so by inserting fingers, multiple slips of paper, and clips within the pages of the book.

Similar problems occur with our personal records. How should we organize a personal notebook? Temporal organization is the easiest, but not necessarily the most efficient. What about address books? We use alphabetical arrangement, but sometimes we want people listed by job title or city of residence or some other characteristic.

Alphabetic organization only works, of course, if the language has an alphabet, and not all languages do. Nonalphabetic languages (for example, Chinese characters or Japanese kanji) have to find other means of organization, usually by finding something that is the equivalent: a well-known structure that can be ordered to aid in the search. The structure need have nothing to do with the meaning of the word, as long as it is systematic enough that people can use it as a search tool. This usually means such schemes as using the order of writing the component strokes of the characters and then organizing the dictionary by strokes, or translating the word into its spoken sounds and then using an alphabet or a syllabary to organize the sounds and syllables.* (Actually, the second way doesn't work well in China, where a symbol may be pronounced differently by different dialects even though its meaning remains the same.) The alphabet simplifies the organization of the dictionary, but it is quite arbitrary, bringing together words that have no other relationship except the coincidence of similar spellings.

How could we restructure the dictionary? There have been many attempts. One example is the thesaurus, a listing of related terms: Look up one word and get a list of others that mean similar or opposite things. Some dictionaries are organized by rhyme, to help the poet or crossword puzzle solver. One novel scheme is now available electronically: Start with the definition and work backward toward the words. The differences in all these schemes are in the ways they provide access. What we really need are intelligent agents: the equivalent of the McGuckin's Hardware Store clerk.

The dictionary, encyclopedia, thesaurus, spelling checker, and language translator all provide similar services, their differences, again, being primarily in the manner by which they are accessed. Right now they are all separate, unrelated applications, but this came about through historical accident. Their function would be greatly enhanced if they could all be combined into one.

An electronic spelling checker is a primitive form of intelligent agent. Type a word and the checker looks it over. If it really is a word, it does nothing. If not, the agent gets to work and provides a set of alternatives, usually based upon both the spelling patterns and known sound confusions. Good correctors are more than specialists in spelling: They also know a bit of grammar (enough to flag repeated words and to suggest capitalizing the starting word of a sentence), and they know about typing errors (enough to change "congitive" into "cognitive") and pronunciation (enough to change "feasant" or "asma" into "pheasant" and "asthma," but not enough to change "kof" into "cough").

The spelling checker provides a very convenient way to get into the dictionary or thesaurus: I type what I am looking for, and it does its best match. But it is flawed. If the corrector is intelligent enough to suggest some spellings, why can't it simultaneously offer to tell me the dictionary definition or the thesaurus entry? And what about grammar? If it knew more grammar, it could help me better. No, I have to call those functions by using different computer programs: Each is specialized for its function and doesn't know about or care to communicate with the others. Uppity specialists. Is this flaw due to a lack of imagination on the part of its creators or to the tangled web of financial and copyright arrangements that surrounds the use of electronic media? Probably both.

Reference works in general are hamstrung by the limits of their technology: the printed book. The book is linear: Page follows page in fixed order. This creates a major organizational problem that cannot be overcome. Sure, publishers and authors try. They invented chapters, a way of dividing up material into smaller, manageable chunks. Headings aid skimming. Tables of contents and indexes are other aids. Each of the aids then has its own organizational problems. Tables of contents are usually stuck at the front of a book, organized in the same order as the book itself. Indexes are usually stuck in the rear, organized alphabetically. These are both arbitrary inventions, however, and their organization and location are equally arbitrary.

One of the virtues of the computer is that it can search huge amounts of material relatively quickly. To aid the search, it might form internal indexes and tables, but the user does not need to know about these. To the user, the computer could present the image of a system that was organized by demand. The dictionary need not have any fixed organization: If I want the definition of a word, I just type it, and *voilà,* there it is. If I knew the definition but not the word, why not let me search through definitions? I could find the word I wanted by searching for its definition, pronunciation, part of speech, language from which it originated, or whatever came to mind. The power of modern technology is that there need be no order: The order could come on command, in whatever fashion the reader requested.

Alphabetical order is such an arbitrary way to organize things that it has always amazed me that we put so much faith in it. Even computer scientists seem trapped by ordering: One of the standard exercises for students is to study the various means of sorting items into order—bubble sort, tree sort, quicksort . . . A wonderful set of exercises, but why? One of the powers of the computer should be that it doesn't have to keep things in order.* The effort to sort things is reminiscent of the fact that at first, computers were used to print accurate mathematical tables of numbers. This effort missed the point of computers: The computer makes it possible not to have to use tables. Do I need to know some number? Just compute it on the spot. Actually, this solution wasn't practical until computers were small and inexpensive enough that everyone could have them whenever needed. This has already happened. Small computers—we call them calculators—are everywhere available. They are cheaper and more convenient to use than the tables they have replaced.

The power of the computer to find information rapidly, regardless of how it is organized, regardless of how it is requested, has already changed the nature of many of our most common cognitive artifacts: reference books. Small, pocket electronic dictionaries (some giving translations in multiple languages), Bibles, and reference books are all easier to use in their electronic version than in the printed one. No more thumbing back and forth through the pages:

Type in what is known of the item of interest—even if misspelled—and there it is. All the statistical information about teams and players for a sport like soccer, football, or baseball can be more conveniently carried and used in a pocket electronic book than in a paper one. My electronic daily calendar and address book, although clumsily designed, is smaller than and superior to the paper versions I used to use.

Electronic media—especially in the case of reference works (tables, statistics, dictionaries, encyclopedias)—can be superior to printed media primarily because they have the ability to solve the search and organizational problem for their users. The schemes of alphabetical organization and extensive indexes currently used by books are unwieldy. Computer-guided search can overcome the limitations of the printed medium, but only if the developers of these systems construct them properly, taking into account the needs and capabilities of the people who use them. Right now it is hit or miss: No devices are yet done properly, some are partially successful, others are failures. But the future path is clear: the potential gain in utility, convenience, and high usability.

McGuckin's Hardware Store shows to what we might aspire: efficient, intelligent agents, coupled with a functional arrangement that makes browsing a pleasure and a source of unexpected finds. The story of Chris's office shows the wide variety of technology available today to help in individual organization. The filing cabinet allows for hierarchical organization. But even it is severely constrained by the limits of the physical structure of the records.

The analysis of the dictionary/thesaurus/encyclopedia shows that new methods are possible. With modern tools, it is perfectly feasible to develop artifacts for maintaining information files *without* any particular ordering. That is, one can store the information internally in any format one wishes but reconfigure it in numerous, flexible ways at the whim of the user. If users have different needs at different times, why not allow the displayed structure (the interface representation) to match the need, not reflect the limitations of the underlying technology?

NAVIGATING THROUGH CYBERSPACE

Ah, the world of computers, invisibly interconnected every which way. My family's computer in the upstairs family room of our home is connected to my computer in my downstairs study, which, in turn, can be connected by telephone to a wide variety of computers. Most of the time it goes to my university computer, which is part of the local campus network, which connects several hundred machines (each a "mainframe," much larger than a personal computer). The university is connected to networks that reach thousands of locations all over the world, perhaps millions of personal computers and people. Cyberspace. A vast interconnected network of information exchange.

Do I need a reference for a book, magazine, or journal article? From my home, I connect to my university library, which has the database of all nine campuses of the University of California. I can look up books, get their call numbers, discover whether each one is available on the shelves and, if not, when it is due back. I can search for journal articles and get the tables of contents of journals and abstracts of articles.

I can connect to Ohio State University and look up articles, books, and technical reports on the interaction of people and computers (human-computer interaction is what the field is called), a database that includes complete abstracts of all the items. If I connect to the University of Michigan, I could find out the longitude and latitude of cities in the United States, organized, so I am told, by postal code. I can even connect to the Scripps Institution of Oceanography pier in La Jolla, California, and find out the water temperature and wave conditions, a favorite and essential service for our local divers and surfers. All of these connections are legal. All are publicly advertised, and people are urged to make use of them. There are hundreds or thousands more, most of which I don't even know about.

With all this information available, with all this knowledge, how can we ever find anything that we need? I have enough trouble finding things in my own office. Look at what Chris had to go through to keep track of her office. With cyberspace—this electronic, invisible collection of information—how will we ever navigate our way to find what we need? How can this stuff be organized? Certainly not in alphabetical order. Where is the equivalent of the clerks at McGuckin's Hardware Store?

The favorite metaphor for those who worry about these problems is that of navigation. People, we are told, are spatial animals with special spatial abilities for navigation and finding things. Certainly, this appears to be true of my life and the way I put things around the house. I know where most things are located. That pile on the upper right of my bookshelf is my unread magazines and journals; the big mound on the bottom left is my "urgent" pile, with all those things that require immediate reply but that I haven't yet gotten around to. That pile in the corner of the kitchen is notes on the work around the house that I plan to do someday, and so on.

My kitchen probably has a thousand different items in it, counting all the cooking and eating utensils, dishes, silverware, the cookbooks, phone lists, recipes, cleaning items, and separate food items in the several pantries. I don't label my kitchen, I simply remember where each thing is: Everything has a proper place.

The importance of spatial information tempts a natural solution to the problem of information overload: Put each piece of information in a different location. Hard to do in real life? Well, that never stopped the inhabitants of cyberspace, which, after all, is not a real place. With computers, we can invent worlds, what computer folks like to call "virtual worlds." We can even display their appearance to you, thanks to the power of computer graphics that can show imaginary places in great detail and with great realism. Just as things are located in my kitchen in their proper places, we could put all the pieces of knowledge available to computer systems into their special places within a virtual world. Look out the virtual window: See the virtual ocean with the virtual pier? Go to the end of the virtual pier, and there is a virtual meter telling you the temperature, period, and height of the waves. The virtual notice board by the virtual meter contains the weather forecast.

Want those longitudes and latitudes? Go back to the virtual house and hop into your virtual flymobile. Fly high enough and you see a virtual display of the whole country, each state nicely outlined, mail codes glowing red. Find the state you want, look inside its virtual image, and there are the longitudes and latitudes.

See how simple it is? Just like the Wooton patent desk—a place for everything, and everything in its place.

The nice part about spatial locations is that when they work, they work well. But what about when they don't work? What about information that doesn't fit nicely into a spatial metaphor? Then spatial metaphors can be frustrating to use, infuriating even. Even in the kitchen, spatial location of all the appliances and utensils doesn't always work. My observations of people in their kitchens reveal that, yes, they all *claim* to know where everything is and that, no, they *really* don't. That is, it is quite common to have to open several drawers or cabinet doors to find something. "I thought you knew where the cheese grater was," I would remark. "Oh, it's supposed to be there, but someone must have moved it," they answer.

The greater the number of people who use the kitchen, the more problems there are apt to be. The larger the kitchen, or the more gadgets owned and stored away in nonstandard places, the more problems. Moreover, there isn't any "standard" location for many kitchen items. Even with appropriate locations, as each location becomes more crowded, search becomes more difficult, and one is apt to miss the item being looked for, even if it is in one of the places searched.

Spatial organization only works when certain conditions are fulfilled:

- There has to be a natural, spatial mapping between the items and the spatial location. The items must have a reason to be placed in their spatial location.
- The knowledge of the locations has to be good enough that the desired items can be located with a minimum of attempts—ideally, one or two.
- The number of different items at any single location should be small enough that they can readily be found, else even if it

were in the correct location, the sought-for item might not be found.

- The amount of work required to try a location, scan its contents, and then try another location should be small.

On the whole, these conditions are apt to be met either for relatively small collections of information that are frequently used or for material that structures itself so nicely that each item does indeed have a logical location, and even if there are many items scattered about, their functional descriptions pretty much determine their place of storage (as in McGuckin's Hardware Store). Searching through a spatially organized structure only works for those who know the structure well, sometimes only for those who constructed it in the first place. Will this scheme really work for all the knowledge of the world? No way.

Through the years, people have tried various means of forming hierarchical organizational schemes that would put each item in its place, once and for all. Then, so the hope went, you could simply navigate through the various organizational structures until the item of interest was discovered, just in its proper location. The problems with these schemes are numerous. For one, there is seldom a unique location for any given item. Where we expect to find something may depend upon what use we intend to make of it. Intend one use, think of it in one way; intend another, think of it in quite a different way. How we think about something determines how we look for it.

A second problem is the assumption that organization can remain fixed over time. Yet the way we think of something today is apt to be very different from the way we think of it later, for as we learn new things, new organizational structures will arise. Couple this to the fact that organizational structures for a society must last decades or centuries, and you see that whole new classifications of knowledge may arise, making obsolete the earlier classification. One need only examine the classification schemes used by modern libraries to see these problems, to see how badly the scheme works for rapidly changing fields.

A final problem with organizational structures has been called "the navigation problem." Consider a computer-controlled database of thousands or even millions of items. As one searches through the database, taking first this path, then that, it is very easy to get lost and confused. Hence the term *the navigation problem*. Actually, the name *the navigation problem* is part of the problem. This invokes a spatial metaphor: Searching a large database is like wandering through a maze of paths and trails, and the guidance required is therefore signs and maps.

What are the alternatives? Simple: Why have any organization? Why have a space? Much as I argued that dictionaries need not be organized but should instead simply make available whatever information you request, why not do the same with the information of the world? Want a good example? Human memory—yours and mine.

Our memories do *not* work by navigation. We do not follow simple paths through memory, aided by signs and maps. No, we think of something, and voom! there we are. That reminds us of something else, and voom! there we are. Actually, sometimes where we are is not where we wanted to be, but that's OK: We just try again. And it is all so automatic and simple that we often are not aware that we are going through a sequence of memory retrievals. Now that's the model I want for our artifacts: so natural that I don't even realize when I have to make several attempts. Or even if I do realize it, I don't care.

I call this process navigation by description. We describe what we care about (mentally, to ourselves), and the system works with that. Any description. The general properties of human memory are that it is easy to return directly to any point we have just been to. Do you remember my description of McGuckin's? See! There you are, remembering McGuckin's. You didn't have to "navigate" to remember this item and that: I simply presented you with "McGuckin's," and there you were.

I'll try again: Think of your telephone number. Out it came, I bet, no search, no navigation: Describe what you want, and there it is. Not only that, but there are no error messages.

Making an error is no big deal. In fact, you don't think of it as an error; you simply didn't get the information you wanted. Was the description incomplete or otherwise inappropriate? Modify it and try again. Note that I said "modify," not "redo"; that is an important difference. You don't have to start over, you simply continue from where you left off.

To summarize, retrieval by description provides alternatives to navigation by pathfinding or by arbitrary orderings, such as by date or letters of the alphabet. Prior to the age of computers, such a development was not possible. Today it is. There is no need to organize material by alphabetical order. There is no need to put it together according to any organizational scheme. Similarly, there is no need to structure information so that there is but a single copy located in a single place. Instead, there can be multiple routes, multiple descriptions, and multiple methods that all get to the same ending point. What matters is that the users be permitted whatever descriptions are most relevant to themselves and that the system accommodate itself to the users.

Today there is still too much emphasis on rigid organizational structures, often devised to save time, money, or equipment rather than to simplify the user's task. Beware of today's new computerized libraries that claim to work by description but have formulated the description in the unnaturalness of mathematical logic. Yes, this is retrieval by description, but in such an unnatural manner that it defeats the purpose. It is precise, logical, and accurate, which is exactly why it does not serve humans well. Our tools must match human capabilities, and these are *not* precise, logical, and accurate.

THE ELECTRONIC LIBRARY

> "I was tinkering around one day and I came across the words *rock salt* in a title," Vander Meulen recalled. "Out of curiosity, I did a search to see how many times the term appeared in the 18th Century. Lo and behold, I found eight pamphlets written in 1760 and discovered what I call the Great Rock Salt Controversy of 1700." (Los Angeles Times, *January 13, 1991, pp. A3 and A29*)

Electronic libraries differ from today's physical libraries in ease of use—an affordance of the technology that facilitates tasks that would otherwise be so hard that they would never get done. The affordances offered by the new electronic access to information databases provide some of the most powerful possibilities for future technologies, along with some of the most dangerous.

Perhaps the most important of technological developments for cognition is the electronic accessibility of a vast compilation of information. Much of the information is not new, much was available before, but in cumbersome, difficult-to-manage form. Birth records were handwritten on certificates filed in courthouses and offices of local cities, towns, and districts throughout the world. So too with marriage records, death certificates, bills of sale and transfer of property. It was possible to trace the history of people only by visiting all the locales where they had lived and done commercial transactions, laboriously fitting together the information culled from countless individual pieces of paper.

The historian, the economist, the government administrator all had to spend considerable time and energy gathering the information relevant to their studies. Today all this has changed. Sitting in an office at work or at home, it is possible to connect to huge databases of information. Ask almost any question and the data are there, somewhere, available to the modern worker. In principle, this precise, efficient availability of information is a boon to anyone who must make significant business or governmental decisions.

There are many reasons to be pleased about this increased access to knowledge. The reason is straightforward: Knowledge is power. Having knowledge available at one's fingertips enables one to do research, to answer questions, to explore new relationships never before possible.

There are also many reasons to be frightened. Again, the reason is straightforward: knowledge is power. If government agencies and businesses have available personal information about the inhabitants of a nation, nothing is private anymore. All a person's deeds, whether proper or not, are scrutinized by hordes of others. The very technology that makes stores more efficient—recording the sale of each item, allowing efficient ordering and display of items, and allowing items to be paid for through credit cards—also allows stores to determine just which customers have purchased which items. Soon, prying people know which newspapers, magazines, and books each of us reads, what type of soap we use or breakfast cereal we eat. From the records of purchases, they can infer intellectual preferences, political party, medical needs, and outside activities. Nothing escapes the notice of the electronic society.

The ease of access to large bodies of knowledge is one of technology's tradeoffs. The positive side is that we can now answer questions that simply couldn't be addressed a little while ago, just as the quotation opening this section illustrates. The quotation describes the work of David Vander Meulen, an expert on eighteenth-century English poet Alexander Pope and a professor at the University of Virginia who now has access to a new computer database that lists roughly five hundred thousand titles published in English from the first printing press in the late 1400s through the end of the eighteenth-century. With the aid of the electronic database, he discovered and studied the "great rock salt controversy of 1700." The search didn't require any information that didn't already exist prior to the electronic record keeping. The new electronic database simply listed the books that have long existed in the libraries of the world. The difference lies in the affordances of the old and new technologies. The computer system transforms a difficult or perhaps an impossible task into one that can be accomplished in a matter of minutes.

The power—and the danger—of electronic databases is that they provide the affordances for compiling information that never before could have been collected without expending enormous amounts of time, energy, and resources. Prior to the computer, the affordances were nil. Records might be scattered all over the world, and to research any given question would take years, enormous energy, and vast sums of money. And whenever a new question would emerge from the search, it would require yet more time and travel to answer. But when everything is on the computer, a question can often be answered at a single sitting. Do the results suggest a new question? Then it may take a second session to answer it. The difference between having all the information readily available and having to travel the world to gather it makes the difference between doing the task and not doing it.

Couple this vast ensemble of data with powerful computers, with portable reading devices, with information dispensers, and you have dramatic changes in our uses of information. Want to know about the restaurants or movies in a city? Insert a personal information reader into the credit card–operated kiosk and transfer the daily schedule into the reader—complete with reviews and travel instructions. With readers and kiosks come new forms of information vendors: vendors that publish electronically. Want to read a book on the airplane? Plug your information reader into the airplane's book dispenser. Want to listen to music, watch a film, practice a second language? All available. How about transferring the daily stock market figures or economic indicators so that you can study them as you fly to your destination? A whole new industry emerges.

The benefits of this ready access to information for education, business, and the individual are enormous. So are the perils. Example: Be wary of that information. Where does it come from? Who writes the restaurant and movie reviews? Will it be the restaurant and theater owners or some independent critics? Who gathers the economic statistics? How biased might the presentation be? Who prepares the educational material? Airline reservation computers are often owned by the airlines. Sure, they contain all the flights offered by all the airlines, but the ones easiest to access (being listed first) turned out to be the ones of the company that owned the reservation system (until the government stepped in and put a stop to the practice). These subtle biases of information sources are often difficult to detect, then difficult to countermand. The company that packages the information in the form that gets it to consumers most easily, in the most usable form, will get most of the business, even if that information is provided by self-serving sources.

There are other dangers. The affordances strike again: It is easy to keep a record of individual actions, including what different people read; what they buy; what stores and shops they enter and what they purchase there; where they live, work, and travel; whom they visit. In part, the system has to know this information in order to bill its customers properly. In part, the information is compiled automatically in the course of collecting routine data about sales and inquiries. "The System" now knows a huge amount about each individual. It knows where you are as well as what you are doing. And with whom. Paul Saffo, writing of what he called "our privacy jitters," pointed out that

> Daily life is nothing but a string of transactions from credit card purchases to phone calls. . . . [E]ach action adds to a spreading electronic wake that we leave behind us as we go through life. A few bits of matched data can tell volumes about us all, a conclusion implicit in the informal motto of the marketing industry: "We know more about you than your mother."

Will society react by accepting the costs along with the benefits? Or will society react to the costs, strengthening privacy by making it more difficult to get access, adding more protections, at the risk of hampering access to legitimate, nonprivate information? Or will it figure out a reasonable balance between protection and free access?

Saffo predicted that it is too late to reverse the cycle:

> [W]e will never slow, much less reverse the growth of this electronic marketplace, because our society cannot function without it. The most that privacy watchdogs can hope to accomplish is to ensure accuracy and confidentiality for certain kinds of information.

Is this so? Is it true that "our society cannot function without it"? Yes, if we accept society as it is, but there is another option: Change society.

EIGHT

PREDICTING THE FUTURE

PREDICTING THE FUTURE IS A POPULAR INDUSTRY. PROPHETS ARE NOT IN short supply. "Quite the contrary," Herbert Simon once pointed out, "almost everything that has happened, and its opposite, has been prophesied. The problem has always been to pick and choose among the embarrassing riches of alternative projected futures; and in this, human societies have not demonstrated any large foresight."

It is easy to devise numerous possible scenarios of future developments, each one, on the face of it, equally likely. The difficult task is to know which will actually take place. In hindsight, it usually seems obvious. When we look back in time, each event seems clearly and logically to follow from previous events. Before the event occurs, however, the number of possibilities seems endless. There are no methods for successful prediction, especially in areas involving complex social and technological changes, where many of the determining factors are not known and, in any event, are certainly not under any single group's control.

Nonetheless, it is essential to work out reasonable scenarios for the future. We do know that new technologies will bring both dividends and problems, especially human, social problems.* The more we try to anticipate these problems, the better we can control them. Even though neither the range of new technologies nor their full implications can be predicted with any certainty, we can make reasonable estimates about the direction of future technology. Attempting to predict the future is both foolhardy and essential. Actually, it is one more thing: It is fun.

EXAMINING THE PAST

A humbling way to begin a look toward the future is to examine the past: Look backward at previous prophecies.* The record is pretty bad. The failures were of all sorts: bad at predicting which new technologies would succeed, bad at predicting which would fail—the overpessimistic seem to be equal in number to the overoptimistic. The most interesting failures, however, involved usage: Even when the technology is predicted properly, it is rare that anyone truly understands its real impact, how it will be used. In fact, I use the word *rare* just to be safe: I have never seen a prediction that was correct about the usage.

In my opinion, the easy part of prediction is the technology. The hard part is the social impact; the effect upon the lives, living patterns, and work habits of people; the impact upon society and culture. I'll lump all of these issues under the term *the social impact of technology.* It is the social impact of technology that is least well understood, least well predicted. That is hardly a surprise, since it is also the social side of technology that is least well supported. After all, the technologists are not social scientists or humanists, they are researchers and engineers. They can be excused for not understanding the social side of their handiwork. However, they cannot be excused for not acknowledging their own lack of understanding and having some social experts join their team.

I start by examining four examples of failure in prediction: the private helicopter, nuclear power, the computer, and the telephone. For each, I'll pay special attention to the social aspects of the prediction and the technology. Then I examine three examples of failure to predict the time frame of technological development: television, the airplane, and facsimile machines. Each of these took a surprisingly long time between its initial invention and widespread adoption.

The Accuracy of Predictions

The private helicopter. The prediction was simple: We would all have our own private helicopters. This was a basic extrapolation of the success of automobiles or private aircraft, and of the new developments in the autogyro and the helicopter. In a helicopter, the rotary wing provides both lift and propulsion. In the autogyro, these two functions were separated. The rotary wing provided lift only: It was freewheeling, rotating without being powered. A normal propeller at the front of the aircraft provided propulsion. Autogyros have disappeared, but in the late 1920s and early 1930s, they were predicted to be the transportation of the future.

The helicopter is a clumsy machine, expensive, inefficient, and noisy. Moreover, it is inherently unstable, unlike the fixed-wing airplane, and therefore difficult to fly. I have flown airplane simulators without any formal training, and although I managed to crash both a 747 and an F-18, I did fly them for considerable distances first. I once tried a helicopter simulator: I lasted about a minute. Increase the power and the helicopter spins in addition to rising and moving in whatever direction the tilt of the rotors propels it toward: The pilot has to compensate for all these interactions.

The problem with the prediction that everyone would use helicopters daily, much as we use automobiles, is that the prediction focused on the technology of manufacture, not the social consequences, the stability problems, the noise, or for that matter, the danger once millions of helicopters started darting this way and that. If you think automobile traffic is dangerous, consider what it would be like in three dimensions, without distinct, marked roads. The problems of today's air traffic control system would seem trivial in comparison to the chaos that would result from buzzing swarms of flitting helicopters.

The prediction failed. We do not have a helicopter/autogyro in every garage. Thank goodness.

Nuclear power. The prediction: too cheap to meter. In the early euphoria over the technology of atomic fission, the emphasis was on the large amount of power that could be generated from a relatively small amount of uranium. The technologists went wild over Einstein's equation $E = mc^2$: c is the speed of light, a large number (and c^2 is huge); m is the mass of the material; and E is the resulting amount of energy. Take a small amount of mass (an ounce or a gram or two), multiply it by the speed of light squared, and you get an enormous amount of energy. Why, so much energy from so little fuel, it was said, that it would cost more to install electric meters and send bills than it would to provide the energy. Just give it away, either free or for a low fixed fee.

The prediction obviously failed. Its proponents overlooked some important factors. First of all, even if the energy were free, it costs millions, perhaps billions, to deliver it to homes and businesses through the massive electric power lines that are needed and that we do have today: A major part of the cost of our electricity pays for the delivery system. Second, the critics neglected the complexity of the plant necessary to harness small amounts of fissionable material. The real flaws in the prediction involved this aspect of the technology. The plants are large and complex. They are designed, constructed, and maintained by people. Technologists always seem to think that people are not human, that they work like machines. But people are not machines, thank goodness. Ask a person to act like a machine and you will be disappointed. Disappointment prevails throughout the nuclear power industry.

It's a social issue. In my opinion, the plants are too complex to be run.* Look, in every large building, there are continual failures of equipment. That's true even in my house: The roof still leaks, despite repeated attempts to cure it, and my kitchen stove has a broken switch. Nothing special about my house; most homes have similar minor failures. In any large building, lights will burn out and not be replaced for days or even weeks, pipes will leak, fans and heaters and blowers cease to function. All part of the day's routine. A nuclear power plant is a large complex of buildings, so it has a normal amount of equipment failures. The problem is that in the nuclear part of the plant, these minor ailments are more critical: The pipes carry high-pressure steam, some of the liquid is radioactive, and a huge amount of heat is being generated, which creates critical energy-balancing problems. The heat can't be dissipated quickly: If the plant has to be shut down, it takes a long time to do so in a safe manner and then a long time to restart again. Some parts of the plant get corroded over the years because of the high pressures, the liquids, the high temperatures, and the radioactivity. These parts are often difficult to inspect and replace.

Nuclear power is simply not safe, but not because it is nuclear. It is dangerous because the plant is too large and cumbersome. I think the same of large chemical processing plants (the number of accidents and deaths from chemical plants exceeds that from nuclear power, buttressing my argument). If the plants were built and maintained exactly as specified in the plans, there might be less of a problem, but in real life, this is an impossibility. Nothing is ever perfect. Here is not the place to discuss the details of nuclear power but simply to point out that the failure of the predictions was really a failure to consider the entire picture, especially the human, social side of the operation.

The computer. In the late 1940s and early 1950s, computers were huge, complex devices, thought to be of limited use: Four or five would do for the United States, three or four would suffice for Great Britain. The failure of the predictions about computers was twofold.

First, there was a failure to predict the invention of the transistor and the resulting semiconductor technology that would miniaturize those early monsters. The first computer I ever used, the Univac I, had one thousand words of memory and was so large that I could (and did) walk inside the central processing unit (the cpu, the thing that today is a tiny plug-in chip). It had several thousand vacuum tubes, which produced a lot of heat and continually failed (if one vacuum tube had an average life of two thousand hours before failing and the computer contained several thousand of them, the machine would fail, on average, every hour). Because everyone assumed computers would always be constructed of vacuum tubes, they could not anticipate personal computers, small machines on desktops that are incredibly more powerful than the Univac.

The second failure was a misunderstanding of what a computer does. To the early technologists, computers computed—

hence the name. And computing means numbers. Today computers manipulate information, not numbers. They create music and art, pictures and sound. Words. Books. Pictures, both moving and fixed. Computers are part communication device, linked across the world, part television set. Computers will replace books, maybe. This was missed completely.

I recall one conversation in the early 1970s with several leading computer scientists in which we tried to figure out why anyone would ever want a computer in the home. "What would the average person do with them?" we asked ourselves. "Games?" "Recipes?" "Income tax?" We laughed and gave up. We failed to predict the networking of computers to the telephone system, their role in entertainment, and their ability to provide access to large amounts of information, as in encyclopedias and other reference works. For that matter, we underestimated games. Never did we imagine specialized computers plugged into the home television set, dedicated to game playing, toys with more computing power and graphic capability than we then had in our laboratories. We underestimated the power of the experiential mode.

Of course, in some sense, we were correct. It is still true that relatively few families purchase computers for their homes. Instead, the computer is sneaking in under other guises: the interactive compact disc, the microprocessor hidden in the TV set, the kitchen appliance, and the automobile. The notion that we would carry computers around in our briefcases or shirt pockets or that delivery people and police officers would use them routinely was never even considered.

The failure to predict the computer revolution was a failure to understand how society would modify the original notion of a computational device into a useful tool for everyday activities. Progress required a different, more enlightened view of the computer as an information processor and controller, where information includes words, sounds, and pictures.

The telephone. Ah, here is where the failure to understand society shows up most strikingly. The telephone was almost immediately recognized as a marvelous invention, but how it might actually be used was not so obvious. Early users often discovered they had nothing to say. Even the enthusiastic were insufficiently imaginative. The problem was to figure out what the phone could be used for. According to an early, enthusiastic prediction, "Why, the telephone is so important, every city will need one!" The idea was that everyone could gather round the phone to hear the day's news. "The telephone can broadcast the daily news to the assembled populace every evening." A version of this idea was actually tried: The Hungarian service *Telefon Hirmondó* transmitted daily programming to more than six thousand subscribers for a number of years prior to World War I.

Trying to determine appropriate uses for the telephone continued for quite a while in the early days. A fire department refused to accept a phoned-in report of a fire because it "was not according to the 'official' routine." The fire department operator "declined to give any alarm and remained quietly in his office for about ten minutes" until an official, proper alarm was finally received. You see, everything has its place in the world. As for the house? It burned down.

At first, the telephone was restricted to the elite. In 1884, when an Edinburgh telephone company proposed making pay telephones available, one person complained, "if everybody who has a penny or threepence to spare can insist on being listened to by any of the leading business establishments of the city, we shall only be able to protect ourselves against triflers and intruders by paying less regard to telephone communications." The problem became so serious that by 1897, people were requesting that their names not be listed in printed directories. In 1898, the Chesapeake and Potomac Telephone Company tried to remove the telephone from a Washington, D.C., hotel because the management allowed mere guests to use the hotel telephone. Today, of course, all hotel guests have access to public telephones, and better hotels are careful to provide every guest with one if not two or three in the rooms.

The point is not that such actions appear incomprehensible to us now, but rather that when new technologies are introduced, especially ones that alter normal social structures, it takes quite a while for society to adapt. Actually, the phrase *for society to adapt* is inaccurate. Both the technology and society adapt. There is a long, slow period of mutual adjustment as the technology, industry, home, and society all gradually adjust, evolving their behavior and structure for mutual accommodation.

On second thought, maybe those predictions were accurate. Today the average household receives a continual barrage of phone calls from solicitors, telephone marketers, machines with prerecorded messages, as well as wrong numbers and crank calls. Rather than being failed predictions, perhaps they demonstrated clever, insightful understanding of the pitfalls of universal access to machines. False then, true today.

The Time Frame of Predictions

Suppose you are absolutely certain that a technology that you know about will succeed. How long will it take to have a major impact? Warning: Technologies take a surprisingly long time to become accepted.

A good way to appreciate the time frame is to think backward the same amount of time that you might wish to predict forward. Do you wish to predict the state of the communication industry ten years from now? Then start by remembering what it was like ten years ago. Surprise: Ten years ago was not that much different from now. So too will ten years from now not be that much different from now. Fifty years, now that makes a difference, but not always as much as you might think. There are countries today that rely on the technologies of fifty years ago. It takes a surprisingly long time for a new technology to have a worldwide impact.

Any large change must be accompanied by massive alteration of the supporting infrastructure. It is the infrastructure and the inertia of the existing technologies and customs that slow down the introduction of new ones. One problem is called "the established base." If a new technology is to supplant an old one, then people must somehow be convinced to give up the old technology. This is not always possible. We live with many outmoded, inefficient technologies because the potential gain does not appear to be worth the pain and cost of change. We still use the "qwerty" arrangement of typewriter keys, even though better organizations exist. The world uses inferior standards for radio and television broadcasting and low-capacity, low-bandwidth telephone lines. Although change will eventually come about, it takes a very long time. Space colonies should be able to come into being quickly with the latest of technological innovations because they have no established base, no traditions or older technology to replace. But an established city or country cannot so rapidly change its ways. It takes a long time for new ideas and new technologies to propagate themselves throughout the world.

Consider how long it took to develop some of today's most important products of technology: television, the airplane, and the fax machine.

Television. The concept was first proposed in the 1880s in France and Germany. A reasonable mechanism, using cathode-ray tubes (which we still use today), was first described by a Scottish engineer in 1908, only one year after regular radio broadcasts started. The first television picture was transmitted in Britain in 1925, the first scheduled television show in 1930. But even as late as 1949, there were only around 1 million sets in use in the United States, which is barely sufficient for a commercial market. So how long did it take to develop television? It all depends upon how you want to define the starting and ending points, but from basic idea to routine household use took about seventy years.

The airplane. Powered flight was conceived and attempted in the late 1800s. The Wright brothers made the first successful powered flight at the end of 1903. There were dozens of manufacturers for years afterward, but flying was considered a sport and not for serious use. The first practical usage didn't come until the late 1910s, and regularly scheduled commercial passenger flights didn't start until the mid-1920s in Europe, somewhat later in the United States.

Facsimile machines. Facsimile is an interesting case because the growth within businesses seems so rapid. Surely this is an exception to the slow growth of technology, right?

Nope. Facsimile was patented in 1843 by the Scottish inventor Alexander Bain. Commercial facsimile services for the transmission of photographs started in Germany in 1907 and in the United States in 1925. Facsimile was widely used by journalists and photographic wire services from the 1930s on, but it did not become widespread in business until popularized by the Japanese in the mid- and late 1980s. Even today, in the 1990s, it is still not common in the home. From conception to practical use by the wire services? Over 60 years. To practical use in business? Over 140 years. To routine usage in the home? Not yet.

The long durations required to move an idea from conception to practical application and acceptance imply a simple rule of thumb for technological innovation: For a technology to be available to the home or business in ten years, there must now be working prototypes within the development laboratory. Not being conceived, not being studied—working. Ten years is not a very long time to develop a robust, reliable technology; determine what form is best suited for use; interest sufficient customers in trying it; and establish the necessary supporting infrastructure of manufacturers, distribution outlets, and maintenance and repair facilities. Few technologies can exist in isolation from a rich, supporting infrastructure. Book publishing requires authors, paper supplies, and distribution systems, plus a public that can read; television requires scriptwriters and actors, camera operators and directors, lighting and sound engineers, studios and transmitting stations, antennas or cable connections, and finally, viewers. New technologies require new supporting staffs, people who have to learn the new technology, experiment with its potential, and produce the large quantity of material required to make the effort successful. It takes a long time for an infrastructure to be put into place. As a result, many worthwhile technologies fail, many less worthy technologies succeed.

New developments take a long time to travel from initial conception to practical acceptance. Many laboratory developments, announced with great excitement, never reach commercial acceptance. I therefore predict that during the next ten years—starting from whenever this sentence is read—there will be no technological surprises in the introduction of new technologies to the marketplace. Everything that happens can be predicted by examining what is taking place now in university, government, and industrial laboratories. Of course, most of what is happening in the laboratories will never reach the stage of practical application, and the ability to figure out ahead of time which ideas will succeed and which will not is very limited. The main surprises will involve those things that do exist in the labs that many think will become practical, important applications but that will never make it.

PREDICTING THE FUTURE

Here are some easy predictions about technologies, predictions widely shared by people in the relevant technological industries:

- An increase in the availability of "digital information," information encoded electronically in a form that is readily stored, transmitted, and displayed. Today there is a variety of media that deal with digital information, but each in a form incompatible with and separate from that used by the others. This will change: The technologies of computers, telephones, television, electronic mail, and facsimile will all merge into one system. Digital information will be stored in vast international databases and local archives, available for purchase on large-capacity storage devices or direct connection through telephone, cable television, or even satellite links. Some of this will be personal information of a sort never before collected together in such accessible form. Business, government, thieves and scoundrels, friends and lovers, and the just plain nosy will be able to learn facts about all of us that many of us would prefer to keep private.
- Widespread availability of high-capacity communication links reaching into every home, and even to every person, via various forms of wired, optical, and wireless transmission, including direct satellite transmission, cable and fiber-optic lines to everywhere, and advances in compression technology that dramatically reduce the amount of information required for reproducing text, sound, and pictures.
- Continually more powerful computational devices, smaller and less expensive than ever before. Information readers,

hand-held devices that substitute for books, with (eventually) the same convenience, plus better search and annotation capabilities.

- The ability to experience three-dimensional sight and sound coupled to a person's own body position and movements. To be used for conferencing, and for exploring new places, whether real or imagined. For educators, designers, explorers, or just for entertainment.

The real issues are political, economic, and social, not technological. I have heard a U.S. congressman state that "every child will have complete access to the entire contents of the Library of Congress." Such a prediction boggles the mind, for it shows a very naive view of the issues. Yes, in principle, it is possible. But in practice? Even imagining that it was feasible to transform all the printed information in the library to machine-accessible form and that the appropriate networks and communication links to the homes existed, there remain two major problems: (1) How would you find anything in all that material (see Chapter 7)? (2) Why would the publisher allow you to get such unlimited access?

The second problem is the hard one: the copyright ownership problem. Today, in most nations, ownership of information is carefully protected by patent, copyright, trademark, or even as a trade secret (all are distinctions used in the world's legal systems). Authors, artists, and publishers expend considerable time, effort, and money to prepare their materials: They expect to be compensated for their efforts.

In Chapter 4 I introduced the notion of "technological affordances," the idea that technologies make some activities possible or easy, other activities impossible or difficult. Activities that are easy to do tend to get done; those that are difficult tend not to get done. This lesson—as pointed out in Chapter 7—clearly applies to electronic libraries. In theory, we could copy unlimited amounts of material from today's printed books; in reality, it is impractical. Electronic libraries will change that. But free, unlimited electronic access to libraries diminishes the ability to compensate the publishers, unless the technology both charges for each access and also prevents transferring the material to the home or office, because once there, multiple copies could be made without detection. Not everyone welcomes this prospect.

Once the issue of ownership of and compensation for information is raised, it often deteriorates into savage debate about the morality of this method or that, who really owns or controls "intellectual matter" as opposed to physical matter, and the different perceptions of fairness to the authors, publishers, and vendors on the one side and the readers, students, and public on the other. I can readily be convinced by both sides of the argument: On the one hand, information should be freely available to all; on the other hand, information should be restricted so that those who created the information can be compensated for their efforts. Seeing both sides of an argument makes it even harder to know what suggestions to make.

Once quick electronic access is available, what will happen to publishers? Will printed books disappear, if only because it will be so easy to get electronic access that an insufficient number of people will want the printed version? Will electronic books take their place? This is a difficult issue, one for which strong predictions and opinions come readily to mind. I cover this in more detail in Chapter 10. In the scientific community, more and more journals are available over the computer networks. A few journals are even now only available electronically. How do they handle the problem of compensation? They bypass it.

Today's access to electronic media is handled in several restricted ways. Commercial databases of newspaper and periodical articles exist, but because the user must connect to special facilities to access them, the suppliers are able to charge for the amount of time the user is connected and for the amount of material retrieved. Some large databases are sold to libraries on magnetic tape or, increasingly, compact discs or other laser-disc technologies. These cost enough that the companies can make a profit by selling only to libraries, even if libraries offer unlimited free access to their patrons. Finally, electronic journals, available through electronic mail, are mainly offered by nonprofit scholarly societies.

Scholars are never paid for their contributions to scholarly journals (it is an obligation and a necessary part of scholarly life to publish), so compensating authors is not an issue. Whether the nonprofit scholarly society can afford the editorial and communication-computational costs is not yet known. Bibles, dictionaries, and other reference works are already available in electronic form. Will this system work for less popular material? How well will these limited experiences transfer to other forms of commercial publication? Good questions.

Want some more predictions? Here are some:

- *Publishing:* The rise of electronic books, magazines, and newspapers. New services delivered to the home over communication media using all sorts of display devices, from those that cover a wall to ones worn on the body, combining printed and displayed text, sound, and video.

  The digital medium is truly different from the media of earlier technologies because there is so much flexibility in the way the digital representations can be presented to the readers. With the print medium, the publisher determines what the information should look like: Once established, the user cannot change it. With digital media, each reader could see a different format, a different selection and organization of the material. Of course, for this to be done well requires some consideration of the people who must use the material. Ah, there's the rub: Want to predict whether the designers of the future will be more successful than those of the past?

  With the technology of digital publishing, people could select only those things that interest them in the preferred format, ignoring the rest. Is this a good idea? Not clear. For the individual, yes. For society, maybe not. John Seely Brown has pointed out that one of the things that binds together a culture is that everyone reads the same newspapers, sees the same shows. That gives people common themes to talk about, a common structure for their lives. With this new technology, there might be no commonality at all.

- *Education:* More computer-assisted tutors. More interesting educational software, more powerful teaching resources for the schools, for industry, and for self-teaching at home. Will these tools also cost far more than we are accustomed to spending today? If so, will their benefits be thought to justify their cost?

  Will these be gimmicks, or will they truly enhance education? My fear is that they will succumb to the pleasures of experiential media at the cost of reflection. I'll come back to this in Chapter 10.

- *Entertainment:* The real benefit of many of the new technologies will be the enhancement of experiential modes of presentation. Every age seems to provide entertainment for the masses, whether through the traveling troubadour, the local storyteller, the development of games and formalized play, or popular pulp books. Now we will have new media: three-dimensional television, audio, and vicarious experiences of all sorts. Never before has entertainment had such power.

  Never before have the entertainment companies also been owned by or been in such close partnership with the computing, communicating, and media technologies. The seductive powers of experiential cognition may be overwhelming: Is this what we want? Especially as these entertainment media are ever more tightly interlinked with databases that contain personal information about the people viewing the presentation, coupled with commercial messages continually extolling this product or that. What will these new, increasingly attractive forms of presentation do to the minds of the viewers?

- *Communication:* What happens when there is videophone for the home, videoconferencing for business? Videophone for the home is like three-dimensional films and video: a compelling idea, continually tried, continually failing. Many of my colleagues in the telephone industry deny that these developments will come to pass, and they point to the multiple failures of experimental trials in the past. I say videophone and videoconferencing will become the standard. My colleagues obviously do not watch enough science fiction shows. Their fears are correct, but insufficient to kill this attractive technology. The major issues are both social and technological, once again. Here is an example of a technological problem that has important social consequences: How can you place the camera so that when you look at the screen, the people with whom you are talking see you looking directly at them? What are the privacy issues? How do you answer the telephone when you are in the bathtub or bedroom? What about the obscene videophone call? New rules of etiquette will arise, indicating when it is proper or improper to allow or to deny the video. New behavioral patterns will emerge. The affordances of the videophone differ from those of the telephone: No longer will it be possible to hold a long telephone conversation while reading a newspaper or working on some unrelated task—the video picture will reveal the duplicity. I predict these issues will all get resolved, the technological ones by the technologists, the others by agreed-upon conventions for both the caller and the receiver.

- *The workplace:* The new technologies support diversified and distributed work, social and educational groups. Given sufficient technology, cognitive tasks can be done as well at a distance as at the same location—with, of course, new problems and limitations. When people work at home, removed from direct supervision, there is room for abuse by all parties to the activity. It is easy for employers to take advantage of the workers, too difficult to regulate health, safety, work hours, and age. For many jobs, it is not possible to know whether an employee has worked all the hours claimed; for others, every second of activity can be monitored. A major problem is the lack of social interaction. Workers who are distributed in space lack the social ties that bond members of groups that are physically together. There is no common mail or coffee room. No easy way to gossip or to turn to a coworker for assistance. Will these problems be overcome?

Problem Areas

Every one of these technological predictions raises the possibility of severe drawbacks in addition to their perceived benefits. Some of these problems have already been discussed. Other problem areas include privacy, societal imbalance of access to the new technologies (the haves versus the have-nots), dealing with society's sociopaths, and the impacts upon social interaction.

Privacy. The most obvious problems involve loss of privacy. This is especially difficult to deal with because the entire notion of what is appropriately private is not understood. The convention of privacy has changed over the years and today varies dramatically across cultures and nations: Europe, Asia, the Middle East, Latin America, the different regions of Africa and North America—all have very different views of privacy. It is hard enough to deal with these issues locally, but what happens when information is freely transferred across national borders? Today some countries prohibit the flow of personal information about their citizens across their borders, but policies vary enormously.

It is bad enough that one's political, racial, sexual, and religious beliefs may be circulated to people and organizations unknown, but what of the possibility that much of what is circulated might be false, either deliberate fabrications designed to discredit or accidental mix-ups and confusions in the information databases? Some of this could come about through government interference, some could even be at the level of personal jealousies: Antonio, hoping to thwart John's growing relationship with Helen, fabricates an electronic message between John and Susan in which John apparently disparages Helen and then arranges that Helen "accidentally" encounters it. This sounds like the plot of an afternoon soap opera, but not only could it come to pass, it already has. How does one prevent and correct these abuses? Some form of governmental agreements and monitoring bodies would seem to be necessary, probably all the way to the international level.

It is already possible to eavesdrop on others through hidden microphones and transmitters. Will this get worse? Already we can monitor people's locations and actions. Some companies have experimented with "active badges": The identification badges worn by employees that usually contain a photograph and identification now become monitors, so that management can tell just where each employee is—who talks with whom, how much time is spent in the toilet, in the coffee room, in the office. Some companies already count the number of keystrokes typed per hour; some listen in on telephone calls and read electronic mail. Where does legitimate supervision by an organization end and privacy begin?

Societal Access to Technology: The Haves Versus the Have-Nots. A second problem is economic: Access to information is expensive. High-quality access—visual, high-resolution images, perhaps with high-quality sound and a large amount of information—requires high-capacity communication channels and high-quality display devices. Collection and dissemination of the information are expensive. Who pays for all this? Whether subsidized by national governments or paid for by each user, on demand, these developments still pose the danger of increasing the gap between the well educated and wealthy and the less educated and poor. Those who are educated and skilled and have the money to afford these new technologies can benefit from them; those without will fall further and further behind. I cannot believe that this is what we desire either.

Sociopaths. And how will we deal with sociopaths, those maladjusted individuals who delight in crank telephone calls, obscene messages, and deliberate break-ins that sometimes damage personal files? Some of this is done by the young, newly empowered with computer and telephone technology, eager to learn and then demonstrate their prowess in getting access to confidential information. This is the result of adolescent experimentation that we observe in all forms of behavior: alcohol and other drugs, careless sex, reckless driving, and petty crime. This is probably manageable through the normal routes of education and patient, continual counseling at home and school. But what about adults who continue with this behavior, sometimes intending only mischief, but often intending harm? The new technologies do not change the existence of these people, but they dramatically enhance the impact of their acts.

As society becomes ever more reliant on information and communication, even a temporary disruption of service or in the accuracy of the information has large repercussions. Imagine if someone intercepted the transmission of stock market data and altered the prices? Or changed or delayed air traffic control information? Today no single individual can manage this. In the future, it will be possible. Even the most banal of break-ins to an information database can result in a huge cost to the information providers in both time and effort, because it is not possible to tell whether the break-in was harmless without checking the integrity of all their information and services.

Personal Interaction. The new technologies are bound to affect personal interactions, but in unknown ways. Will we become less social, preferring the safe harbor of machine-mediated interaction, perhaps where we can appear to others in forms different from our normal selves? (I discuss this later in this chapter.) Will we work more at home, less in the office, and if so, how will that affect interpersonal relationships and the working environment? And what of those who are incapable of using these technologies because of lack of money, education, requisite skills, or even by preference? Are they apt to be disenfranchised?

THE POWER TO FANTASIZE

Given the predictions about basic technologies, what difference will they really make? It is one thing to imagine electronic access to books at home or with portable players at the beach, but these seem like a natural progression from existing libraries and books. Many of the basic predictions have that air about them—obvious modifications of what already exists. But what about the radical new opportunities created by the new technologies? Let's look at some exercises in fantasy.

Want to compose some music? Let the technology guide your composition and performance. The same with painting and art. Want to visit ancient Greece or China? Experience those times and places in three-dimensional color, three-dimensional sound. This set of predictions is intended to be stimulating, but still well within the realm of possibility. All could be done within the next twenty years—if not in the home, certainly within the laboratories. No predictions of extrasensory phenomena, no predictions of brain taps and mind reading. No direct mind-to-mind communication. Only things that are possible.

Not all these predictions lead to desirable outcomes. "Blasphemous," one reader of a draft of this book responded upon encountering the statement "let the technology guide your composition and performance." But blasphemous or not, the prediction still stands. In fact, that particular prediction is very likely to come about sooner rather than later. Nonetheless, the fact that technology can do things does not mean it should. What happens to art and music when every person can achieve flawless production? Not to worry, say some, we will still have the unique human ability to create, to have an artistic sensibility: The performance may be flawless, but the essential human part of deciding what to do, how to arrange it, will still be there. Maybe. I would like to believe that. But we already have at least one computer program that generates art of quality high enough to be exhibited in major museums around the world.* This particular program, named Aaron, owes its artistic abilities to the rules embedded within it by its creator, Harold Cohen, a major artist in his own right. Cohen insists the paintings are the computer's, not his: He wanted them to be signed "Aaron."

Start off with the technologies that allow one to visualize being in some make-believe, artificial place. Use a visual display that presents a high-quality image to each eye in order to produce three-dimensional vision; use an auditory display that creates realistic, three-dimensional sound. Perhaps even put on gloves or body suits that both sense the body's movements and location and also provide complete sensory feedback. If some of the newer technological experiments work, then it will not even be necessary to wear special clothes or devices (although you will be restricted to a specified volume of space). These are the technologies being explored in laboratories around the world:

"virtual reality" or "virtual presence," they are called. (Computer scientists love the word *virtual*, by which they mean something that is created so as to have the appearance and behavior of the real thing but that is only a representation that mimics the critical properties.) It is already possible to create computer-generated three-dimensional images of planned new buildings, the goal being to allow architects and clients to "walk" through proposed buildings and experience them from within and without. These technologies allow someone to try using a facility before it is constructed, while it is still easy to make changes. Related technologies can allow chemists to study chemical reactions by manipulating images of molecules, not only seeing how they fit but feeling the force fields through tactile feedback to the manipulators that they use to move the molecules. Today this can only be done in crude ways, but technologists have hopes of great advances.

Will these systems work? Of what use will they be? One success story is airplane flight simulators. The high fidelity and accuracy of these simulators make them indistinguishable to the average person from a real airplane flight. They are accurate enough that aviation authorities allow a pilot to complete flight certification requirements by training on simulators instead of real aircraft (which are more expensive and less forgiving of errors during training). This is virtual reality at its best. But the simulators are not cheap: They cost millions of dollars.

Simulated experiences have the potential to become powerful instruments of cognition. They support both experiential and reflective processes: experiential because one can simply sit back and experience the sights, sounds, and motion; reflective because simulators make possible experimentation with and study of actions that would be too expensive to try in real life.

I have seen medical simulators that allow physicians the opportunity to experiment and learn new procedures, with clever models of patients coupled with videodisc presentation of the same sights and sounds as would be experienced with real patients. These are reflective tools because they allow one to compare and contrast simulated experiences and imagined ideas. The comparisons can then yield improved and enhanced new ideas, new conceptions. There is great educational potential as well.

If we now simulate the experiences of other places, whether real or imaginary, what happens when we can simulate people, or at least their appearances? It is already possible to generate computer-controlled images of places and moving objects that are almost indistinguishable from real images. It isn't yet possible to simulate animate objects with such fidelity, but that day will arrive. Suppose that day is now, and furthermore, suppose we have created devices that allow us to simulate our own appearance, then what?

Faking one's abilities is a time-honored procedure among thieves and scoundrels. The lesser version of presenting a false personality is common even among everyday folk: Witness the popularity of climbing and biking clothes among nonclimbers and nonbikers, or fancy running shoes and apparel among the sedentary, or baseball caps among nonathletes. Why, you can even purchase fake automobile telephone antennas to make your car look as if it has a cellular phone even if you can't afford one.

Imagine raising these tendencies to new heights. Authors and actors have long known that the persona they project through their works can differ significantly from their true personality. Suppose some people use this phenomenon to their advantage. Highly talented letter writers can project any image they wish, as long as there is no danger that they will ever meet their correspondent. A "well-regarded female correspondent" in an electronic discussion group actually turned out to be a male psychiatrist, creating considerable uproar among the other group members when the deception was discovered. How many others mask their true identity in electronic interactions? In 1984, Vernor Vinge wrote "True Names," a science fiction story of a world in which people invented artificial personas, artificial existences with which to interact, while vigorously guarding their real identities and true names.* You went off through cyberspace, exploring the pathways and information, encountering others along the way. Except the others that you saw were also artificial personas. That fierce, cruel sadist might in real life be a timid, prim, and proper old lady. The names were false as well, and it was not considered proper to learn the true identities or true names of the people whom you encountered. How much longer before that fantasy becomes reality?

Why restrict our artificial personas to other people? Why not use them for our own pleasure as well? Mirrors today still rely on the ancient method of reflecting technologies: metal deposited on the back side of glass. Mirrors are invaluable, but they have limitations. We have to stand "just so" in order to see ourselves, and because a mirror image appears to be reversed left to right, we can never see ourselves as others see us. Moreover, with a mirror, we can't easily see ourselves from the side or from behind or above. And it is quite impossible to see ourselves in a single mirror with our eyes not looking straight into the mirror.

But consider the video mirror, a mirror that is really a video presentation from a video camera. Now the image can be taken from any angle, in any size. It could be a mirror-reversed image or a true image (where the right-hand side is on the right side of the image's body). Real mirrors show you as you are now, at the instant of viewing. Video mirrors do not have this restriction: Arrange yourself and, when the image looks right, freeze it—or capture a ten-second fragment—and then contemplate the mirrored representation at leisure. Now that's true reflective thought (pun intended), even if for a less worthy motive than the enhancement of the world's knowledge.

Want to decide between two sets of clothing? Why not use the video mirror to save two different images, one of you wearing each outfit? Then compare the two pictures, perhaps side by side. In fact, why should you have to put on the clothes? Let the computer inside the video mirror "paint" the clothes over your image. The video mirror would enable us to do operations that are otherwise quite impossible. Note that this is true reflective comparison: The video mirror allows comparison of representations. Without the aid of the video comparing mirror, the comparison had to be done in the imagination or by other people.

Now imagine a similar activity involving hairstyle or clothing: Want to try a new hairstyle? Project it onto your facial image. A clothing store could let you test new clothes without trying them on: Have the intelligent video mirror arrange the clothes over the body. Do you like the style of the clothes but not the color? A different color is a simple twist of a knob away. While we are at it, why not enhance the image? Let the video mirror make you appear slightly thinner than reality, or let it eliminate that unsightly stomach bulge, straighten your teeth. The image could be a graceful enhancement of your true self, modified in whatever way was desired.

In fact, the image could be of anything. A man could turn into a woman or a child or a dragon or a colorful pattern of light. Couple the artificially enhanced image with an artificially enhanced voice and, for that matter, perhaps even artificially generated words. Where is reality?

It is frequently said that people will fear video telephones because they may not always want others to see them. After all, suppose you are in the bathtub when the videophone rings or, umm, sitting on the toilet or making love or picking your nose or . . . I have always thought this overdone, for after all, one need not be visible to the camera, or for that matter, one can simply have a photograph in front of the camera. But when my flights of fancy about the video mirror come to pass (not *if*, you may note, *when*—these developments will all happen; the only questions are how soon and how expensive), then the image need have only the slightest connection with reality. In fact, you could project the image of yourself in the bathtub, whereas in reality you were hard at work in your home office. The real problem is not that others might see us as we really look at the moment but rather that others might never be able to discover us as we really are.

Are you dismayed by the possibility of such mass deception? Perhaps these artificial systems actually let people project their true inner selves instead of being restricted to surface features. Do you think of yourself as smooth and accomplished, with a sexy, rhythmic voice, but, alas, your "true" voice never comes out that way? With the appropriate artificial persona, it can be so. So what if the video mirror makes a person appear taller and slimmer, more muscular and more athletic, and ten years older (or younger)? Maybe these enhanced images are projecting the "true" person. After all, if that is how people think of themselves, perhaps that is how they really are, and not the false image that their tinny, accented, stuttering voice conveys. Who is to say which is truer, the flaws of reality or the perfection of one's wishes and dreams?

Artificial personas present a vast range of possibilities. The Japanese have invented a whole culture of machines that enhance the voice. They use it for singing, but it could be applied to a wide variety of situations. Sing into the Karaoke system and your voice is enriched, filled out, and accompanied by a professional orchestra. Today's systems are limited. They use prerecorded music and require the performer to synchronize their vocalizing with the recording. But modern technologies will allow artificial intelligences to come into play. The future Karaoke equivalents will generate music on the spot that will harmonize with the singer. The performance will follow the singer, not the singer the performance. In fact, the system will use the singer's voice only to ensure that the artificially produced sound matches the content, rhythm, and duration of the original—but the sound of the singer's voice could be manipulated so as to appear any way the singer prefers. To make the singing always be on pitch, in tune, is the easy part. But the voice could also be sexy or not, weak or strong, questioning or authoritative, dominating or submissive, teasing or honest—whatever image the person wishes to project.

We don't need to wait for cyberspace to encounter the scenario described in "True Names": We can start with the telephone. Already people experiment with the voice message that greets the caller on telephone answering machines. This is so important for some that they purchase professionally made tapes to use for their "greetings." Why not have a speech transformer that changes the words into whatever form one wishes? Perhaps today the voice would be that of a young woman, tomorrow an old man. Imagine the same when videophones become commonplace: How will you know that the image and voice you see and hear are really the image and voice of the person you are speaking to?

Want to play a musical instrument? Fine, just choose your instrument. Blow into the clarinet, play the keyboard, strum the guitar, or bow the violin. Whatever instrument you play, the sound will come from the computer—your sound, just as you wished it to be. Do you really play off-key? Do you have trouble with those fast passages, fingers or lips getting all tied up? Don't worry, the computer will realize what you are trying to do and make the sounds come out just as you intended, maybe even better.

Want some accompaniment? Want to conduct an orchestra or band? Why not? The artificial sound generator can produce any instrumental sound imaginable. Already we have electronic instruments that play accompaniments automatically. The new ones will perform the melody line, the solo instruments, the accompaniments, the voices—whatever parts and in whatever fashion one wants.

If you hate writing, well, we have something for you too. You can already buy sample letters: Want to send a love letter or a threatening business letter? Simply choose one from your favorite word-processing kit. In the future, the system will be able to generate the text from key phrases given to it. Today we can already do that with greeting cards, tomorrow with personal letters.

In principle, is this so different from the hiring of a scribe? Illiterate would-be message senders hired the village scribe to produce letters. Or take the fictional plot, undoubtedly based in reality, of certain characters convincing others to pose in their stead. In the future, it could be done by machine.

What will become of real singers or real musicians? Or authors? Worse, what if the recipient of your letter doesn't like to read and so gives it to a reading machine instead? The letters get written by machine, read by machine: No need for a human intermediary. Maybe that is how we get out of this whole mess. While our machines are all busy reading and writing to one another, playing music, and calling and answering one another's videophones, the real people could go off to the side and enjoy life. Let the machines do the dull tasks of the world. Let the people then take advantage of the freedom to do some human tasks.

Science Fiction and the Technologies for Interacting with Machines*

Perhaps the best way to think about the future is to invent a detailed, comprehensive scenario: in other words, to write a story. Stories have the virtue of compelling the writer to think through the consequences: They force a complete examination and detailed analysis. It's what we in the design business call "prototyping": trying to piece together every little detail so that you really and truly understand just how the technology or society will work. Taken in this light, science fiction is a valuable exercise. Scientific even.

The problem is contemporary science fiction, especially the genre called cyberpunk, has gone wild. The hero plugs the computer directly into the brain. Sure. Let's go back and look at the technology. How possible are these ideas? The problem with postulating the ability to plug a computer directly into the brain is that it is all done by magic, not by any specification of how it is actually carried out. Could we really connect a machine to the nerve fibers of the brain? Is it necessary?

There are hints of other ways. One of the most original notions is in some sense the simplest. Vernor Vinge, in "True Names," proposes that half the interface is provided by imagination:

> He powered up his processors, settled back in his favorite chair, and carefully attached the Portal's five sucker electrodes to his scalp. For long minutes nothing happened: a certain amount of self-denial—or at least self-hypnosis—was necessary to make the ascent. Some experts recommended drugs or sensory isolation to heighten the user's sensitivity to the faint, ambiguous signals that could be read from the Portal. *(Vinge, 1984, p. 14)*

Just as the words of the book provide a rich, compelling fantasy of the experience you are reading about—with a fairly low-bandwidth interaction with the text, I must emphasize—so too could the computer interface provide a rich and compelling fantasy, with equally low bandwidth:

> You might think that to convey the full sense imagery of the swamp, some immense bandwidth would be necessary. In fact, that was not so. . . . A typical Portal link was around fifty thousand baud, far narrower than even a flat video channel.

> Mr. Slippery could feel the damp seeping through his leather boots, could feel the sweat starting on his skin even in the cold air, but this was the response of Mr. Slippery's imagination and subconscious to the cues that were actually being presented through the Portal's electrodes. . . . Even a poor writer—if he has a sympathetic reader and engaging plot—can evoke complete internal imagery with a few dozen words of description. The difference now is that the imagery has interactive significance, just as sensations in the real world do. *(Vinge, 1984, pp. 16–17)*

Does interaction with technology always have to be low bandwidth? No. Here is David Zindell's description of reading in his epic *Neverness:*

> I was confused because I had always used the word "read" in a different, broader context. One "reads" the weather patterns from the drifting clouds. . . . Then I remembered certain professionals practiced the art of reading, as did the citizens of the more backwards worlds. . . . I supposed that one could read words as well as say them. But how inefficient it all seemed! I pitied the ancients who did not know how to encode information into ideoplasts and directly superscribe the various sense and cognitive centers of the brain. . . . To read with the eyes; it's so . . . clumsy. *(Zindell, 1989, pp. 52–53)*

I used to scoff at the notion of providing a plug from machine to brain. However, I have had second thoughts: There may be something to it after all. Consider the nature of human language. On the one hand, it is a rich, complex, and very sophisticated data-communication medium. On the other hand, it is made possible by extremely arbitrary actions of the body: We puff air in and out of the vocal chords, which tense and relax appropriately, opening and closing various nasal and oral passages and manipulating the tongue, lips, and jaw. To speak, we had to learn rather arbitrary but complex muscle controls. The same can be said for sign language:

just as effective and rich as spoken speech, and just as arbitrary a set of muscle motions.

If you examine human skills, you see that we are capable of learning to produce and to encode a wide variety of arbitrary and complex muscle actions and sensory events. It may take years of practice, but we can learn to do amazing things. Look at typing, another very arbitrary skill. It takes months to learn to touch-type, years of practice to get to speeds on the order of magnitude of one hundred words per minute: That's a respectable speed. Consider playing a musical instrument: A pianist playing a Chopin nocturne may be required to play twenty-five notes each second. In twenty minutes, a pianist may play some ten thousand notes, each synchronized, each correctly timed, each exactly the right one of the eighty-eight notes on the piano. Reading is equally arbitrary: Those arbitrary printed shapes on the page represent spoken words. Reading is not an easy skill to master: It takes years of practice. But after mastering the skill, we can read hundreds of words a minute; in fact, we can read faster than we can listen.

Maybe we could learn yet another arbitrary means of communicating ideas, this time with a computer or directly to the brains of other people. Suppose we tapped a fast, high-bandwidth nerve channel. The brain is pretty hard to get into, encapsulated as it is by the skull, but we might be able to connect an electrical cable to the auditory nerve. Or maybe we could tap into the spinal chord or into the nerves that go to and from a hand. Suppose we hooked up a high-bandwidth channel that sent and received neural impulses through this tap. At first, they would simply lead to peculiar sensations and jerky, uncontrolled movements of the body: weird tingles and spasms with no meaning, no coherence. But I suspect that if we undertook daily training exercises—a few hours a day for, oh, ten years—who is to say that we couldn't train ourselves to communicate? It's not clear to me that this task would be any harder than other things we now train ourselves to do.

Note several things about this suggestion, however:

- It in no way re-creates the sensory experience that now exists—it provides a completely different kind of experience.

- It will require extensive training—years or even decades—but then again, it takes two years to learn a foreign language well enough to communicate, a decade to be comfortable. Thcsc timcs are not out of line.
- It might have to start with children, while the brain is still biologically maleable. So only children who were wired up in the first few years (months?) of life would learn this mode of communication. Certainly, it would have to be done before the brain gels, at puberty. Adults would have trouble. They could perhaps pick up a smattering of the interaction, but always with the equivalent of a strong accent.

MULTIPLE MINDS

So far, most of this discussion has focused on technology and the individual, but perhaps the greatest gains are to be expected in the development of tools for social cooperation. When groups gather to work together, their interactions are both aided and constrained by technology. Chalkboards allow the work of one to be visible to all. Several people can work at one board at the same time, but if they try to work on the same portion, their hands and bodies interfere with one another. Several new experiments have allowed joint work on the same surface through the use of televised images of the individuals. Suppose we each have our own individual work surface, whether a chalkboard, a pad of paper, or a computer screen. Each can work unhindered by the others because only each individual has physical access. But suppose we electronically superimpose everyone's efforts? We can now see everyone else's work superimposed over our own. There are several ways of accomplishing this feat, including capturing the individual work by both television and computer, electronically superimposing all the individual images into one, and then projecting the combined image onto each individual's work surface. In one demonstration study, an instructor of Japanese calligraphy was able to place his hand directly over the image of the pupil's hand and move through the proper motions. The pupil, who saw the instructor's hand superimposed over her own, was given a form of guidance and coaching not possible without the technology.

This is the promise of new technologies for cognitive artifacts: new forms of representational devices that make possible social interactions that were not even thought of before the advent of the technology. These technologies promise enhanced interaction, whether the people are grouped together in the same room or separated by thousands of miles.

A wide variety of tools for assisting joint work are now under exploration within a new scientific field called CSCW: computer-supported cooperative work.* The number of applications seems endless, from joint writing and editing of manuscripts to joint problem solving and decision making. Electronic meetings take place over cyberspace, participants communicating via telephone, video, and computer. Some experiments have even involved joint work teams separated by thousands of miles, but with continual video connection between sites so that one could always walk by the video room and say "hi" to a distant coworker just as easily as with other coworkers in the same building (even more easily, the proponents claim). You know how people walk down the halls and see other people in their offices, stopping now and then to chat or discuss an important issue? Why not do the same electronically, scanning offices briefly, greeting the occupants, having impromptu discussions about the problems in front of you. Privacy? Yes, that is an issue, but so far the demonstration schemes have allowed individual offices to "close the door" on the video in much the same manner as one can choose to leave an office door open or shut. When video visits are OK, the video channel is left "open"; when privacy is wanted, the channel is "shut."

But beware of the pitfalls. It is extremely difficult to devise cognitive aids that work smoothly within a group without destroying some of the power of the group. Technology tends to be unyielding, demanding, coercive. Social groups require flexibility, cooperation, and resilience, allowing diverse personalities, interests, and work styles to interact. When technology supports an individual, the individual can still control the situation, and the two, person and technology, can find some graceful means of interaction. When people work together, social tensions can easily arise, avoided only through the goodwill and cooperative attitude of the individuals involved. But add some inflexible technology to the mix and difficulties can soon appear.

Beware of the quick fix, of the natural desires for a technology that will solve problems, sometimes even when the problems don't really exist. Wouldn't it be nice if we could all enter our schedules into the computer and then it could quickly find convenient times for our meetings? Wouldn't it be nice if we could all comment on this joint draft and then just pick and choose the best or merge the ideas? Wouldn't it be nice if we could all have our computers with us in the conference room? Then the report would be finished when we left, the minutes already written, and we could all share the materials or do joint calculations together. Wouldn't it be nice if we didn't have to travel long distances to meet all the time but could interact through computers, video, and voice?

So far, success eludes us. Computerized scheduling? Lots of things are more easily kept in the head: a tentative set of dates for an appointment, a desire to leave early for a family matter, a need to get to a store during working hours, a secret love affair. A meeting scheduler works only if we all keep our schedule up to date, in computer-accessible form. But usually, only managers have up-to-date calendars, often kept by a secretary, and workers neither have them nor want them. Here is a social tradeoff: Those who benefit most from a technology and those who must do the work to make it function are different people.* No wonder these things are continually introduced, no wonder they continually fail. The interests of the parties who must cooperate are very different.

Tools that assist joint work have special requirements beyond those that apply to tools for the individual. They must accommodate group structure and interests, a much more complex task than for individuals. No single technology or method will provide the answer. Different groups will prefer different methods, depending upon the individuals who comprise them, their experience, their national culture, and the philosophy of the organization in which they are working. The technological support of group work and social interaction serves as a prototype for all our hopes and fears for future technology. The new technologies can lead to distributed social groups, cooperating at a distance even better than could be done when they are physically together. It can also lead to a technological hell, with rigid restrictions, continual monitoring, and a lack of privacy and identity. Which is it to be? The affordances built into the technologies may determine the answer.

WHY IS IT MORE FUN TO READ ABOUT THE NEW TECHNOLOGIES THAN TO USE THEM?

Forget, for the moment, your personal qualms about these technologies. Forget the ethical issues, forget the morality. Even forget the feasibility. Instead, concentrate upon the experience itself, the sense of fun that you can get by imagining them briefly. The question I want to ask is, would they really be that exciting and interesting? My experiences have been rather disappointing. Most real applications I have dealt with have not lived up to their potential, even allowing for the primitive capabilities of today's technology. It seems more fun to read about the new technologies than to use them.

Why? Because imagination is more agile than reality. Imagination doesn't have all the burps and glitches, delays and clumsiness of reality. And my imagination is in full three-dimensional sound and vision, with full color, with no grain or noise. My mind has no physical limits. The multimedia, hyper–information space provided to me by my imaginative mind is superior to any possible reality. In my mind, I make no errors, have no doubts or confusions, and always find exactly what I want. Not so with real technology.

I read about new technology that will let me examine a reference book and expand upon references and unknown terms. I read about the flexibility of instructional systems that will show me how to do this and that. And virtual realities that will let me experience new worlds without ever leaving my living room. Music appreciation, art, foreign languages: all at my joystick, or maybe even at the pointing of a finger or the voicing of a wish.

In reality, I oftentimes can't figure out which part of the display I am supposed to manipulate and what is a piece of screen artistry. Or where I am. When I am trying to find any particular piece of information, it takes forever. Often I just want to experience the damn thing, thank you, and not make all those decisions. And most of it is pretty slow and tedious. Reading stuff on a computer screen is simply not as easy as reading it in a book—not yet, anyway. As for reading stuff filtered through the limited resolution of television, forget it.

When I want pure experience, these new "interactive" media force me into reflective mode, force me to read this and that, force me to make decisions at every turn, all of which distracts from the experience. On the other hand, when I want reflection, I either get a Hollywood travelogue with pretty pictures, rich music, and a deadening narrator—pure experiential mode—or a limited amount of elementary text, written for the filmmaker's legendary sixth-grade audience. When I want to look up technical information, I want *encyclopedia*-level depth and accuracy. Instead, I get a politician's cue-card summaries.

What is the problem? The producers of the material (called "software") haven't yet learned the literary genre: how information should be presented or, for that matter, what information and in what form. The current systems seem to be produced either by computer programmers or by filmmakers. Programmers usually have no training or experience in understanding the needs of people. Their expertise is in the technology, so it is only natural that they should try to demonstrate its powers and features. But they are unlikely to know how to tell a story or present deep, reflective material properly. Filmmakers know how to tell a story: They are the masters of experiential mode. But filmmakers are so worried about losing their audience's attention that they feel they must continually bombard it with new experiences. No time for reflection. Worse, no topic has any depth, and the narration—well, best left unsaid. Reflective mode must seem frightening to these people.

These artifacts of fantasy primarily exploit perceptual images. They can provide wonderful experiences, but will they truly enhance our cognitive abilities, or will they mainly be artifacts of experience, for entertainment, for enjoyment? Today the systems are still primitive, but that can't explain all of the problems.

Entertainment is an important part of our lives, so I do not wish to deny the value or importance of the development of an entirely new domain of entertainment. Nonetheless, until we learn how to provide stable external representations that can be examined, contrasted, and transformed into higher-order, more powerful representations, these new technologies will remain devices of exploration and entertainment and fail in their power to enhance cognition. Experiential or reflective, that is the question. Personally, I see great potential in these technologies—potential that is, as yet, unrealized even within the research laboratories.

NINE

SOFT AND HARD TECHNOLOGY

CONSIDER THE FREE THROW IN BASKETBALL. THE PLAYER STANDS A FIXED distance from the hoop and is given one or two chances to toss the ball through the hoop, unimpeded by the other players. All the player has to do is look at the basket and throw the ball into it.

What is it about the free throw that is hard for the player? Throwing the ball with the necessary accuracy. To become good at the free throw requires continual practice and concentration. Amateurs will miss frequently. Even professionals will occasionally miss. What is easy about the free throw? Many things, none of which are even talked about because they are so easy. For example, seeing the hoop. It would never occur to anyone to spend time practicing how to see the hoop.

What would be hard for a machine? Seeing the hoop. What would be easy for the machine? Throwing the ball. If the machine could only figure out where the hoop was, tossing the ball into it would be trivial, a simple matter of computing the appropriate trajectory and applying the required forces. The mathematics would be easy, the perception would be hard.

Now consider a second example. Yoko is showing slides, but the projector beam is too low. She looks at the corner of the room and says to her friend, "Over there." Her friend goes to the table in the corner of the room, gets a book, and brings it to Yoko, who puts it under the front of the projector and proceeds with the slide show.

What is easy for the person? All of the above: realizing the nature of the problem, finding an unconventional use for an existing artifact, using it appropriately, asking someone to help. For the friend, it was easy to understand what was needed, and fetching the book was simple—so simple that normally we wouldn't even talk about it.

What would be hard for a machine, even an intelligent robot? First of all, noticing that Yoko had a problem and that it should help, even without being asked. This requires some empathy with Yoko and with her task. Empathy: That's going to be a hard trait to build into machines, even the most artificially intelligent of them. If the robot were acting as Yoko's assistant, there is no way that the look followed by the words "over there" would be understood. It isn't even a sentence! Where is the verb? What is the command? What should be done? And even supposing the command were precise: "Please go over to the red book with the three glasses of wine on it that is lying on its side on the small wooden table in the southeast corner of the room. Bring the book here and give it to me," then what? The hard part for the robot would be moving to the proper table without knocking something over along the way, managing to pick up the book properly (after dealing with the three glasses of wine), and so on.

There is an easy solution for a robot, if it ever managed to understand the problem with the slide projector: It could just pick up the relevant side, lift it to the proper height, and then hold the projector for the next hour or two during the show. Can you imagine a person doing that?

The things we are good at are the things natural to humankind. The things we are bad at are unnatural. And guess what? We can build machines that perform flawlessly many of the things we are bad at. As for the things we are good at, it is very difficult, today usually impossible, to build machines that can do them. "Why that's wonderful," you should be saying. "What a marvelous match to our abilities! Between us and our machines, we could accomplish anything, for the one complements the other. People are good at the creative side and at interpreting ambiguous situations. Machines are good at precise and reliable operation."

Hah! That isn't what has happened. Instead, technology has decided that machines have certain needs and that humans are required to fulfill them. The things we are good at, those natural abilities, are hardly noticed. Machines need precise, accurate control and information. No matter that this is what people are bad at providing, if this is what machines need, this is what people must provide. We tailor our jobs to meet the needs of machines.

What are people good at? Language and art, music and poetry. Creativity. Invention. Changing, varying the manner of doing a task. Adapting to changing circumstances. Inventing new tools. Thinking of the problem in the first place. Seeing. Moving. Hearing, touching, smelling, feeling. Every one of these things is hard for a machine. Enjoying life. Perceiving the world. Exploiting taste (food), smell (flowers), feelings (amusement parks), body motions (sports). Aesthetics. Emotions such as joy and love and hope and excitement. And humor and wonder. But these are not the humans that technology sees.

It's not that technologists don't care about people. They do—after all, they are people too. The problem arises from the focus upon machine performance. When there is an industrial accident, review teams pore over the site, looking for signs of equipment failure. If none is found, the blame is put on the humans. Thus, in 75 percent of commercial aviation accidents, the blame is placed on the pilots. Human error. People, as we know, are distractible and imprecise. They make errors in remembering things, in doing things they shouldn't have or failing to do things they should have. As soon as one takes the machine-centered point of view, everything automatically leads to a focus upon human weaknesses rather than strengths.

The industrialized, machine-centered view of the person includes such terms as *imprecise, sloppy, distractible, emotional,* and *illogical.* Think about those terms: each one a negative attribution of people, especially as compared with machines. Machines are precise, people are vague. Machines are neat and orderly, people are sloppy. Machines concentrate on their task, but people can be distracted. Humans are emotional, machines are logical. See what a badly engineered piece of machinery we are?

The question is, how did we manage to let ourselves be judged by these machine-based standards? Sure, every claim against us is correct, but who cares? Those are machine-centered concerns, not human-centered ones. If we take a human-centered approach, we get a different characterization of the same five attributes: creative, compliant, attentive to change, resourceful, and able to take a variety of circumstances into account. Liabilities from one point of view turn into assets from the other.

Look at the following table. It compares the human-centered and machine-centered perspectives for viewing people and machines. Each row in the table describes the same attribute for both people and machines:

The Human-Centered View		**The Machine-Centered View**	
People	*Machines*	*People*	*Machines*
Creative.	Dumb.	Vague.	Precise.
Compliant.	Rigid.	Disorganized.	Orderly.
Attentive to change.	Insensitive to change.	Distractible.	Undistractible.
Resourceful.	Unimaginative.	Emotional.	Unemotional.
Decisions are flexible because they are based upon qualitative as well as quantitative assessment, modified by the special circumstances and context.	Decisions are consistent because they are based upon quantitative evaluation of numerically specified, context-free variables.	Illogical.	Logical.

Under the human-centered view, people are superior, machines inferior. Under the machine-centered view, machines are superior. Which point of view should we be governed by?

MACHINE-CENTERED VERSUS HUMAN-CENTERED VIEWS

We owe much of our intelligence to artifacts, to the technologies of information acquisition, storage, transformation, and display. But even though technology has been of great benefit, at times it feels as if we live under an uneasy truce, with technology tending to dominate. The technology did not create itself, it has all come from us—conceived, invented, constructed, and applied by us, for us. Nonetheless, the needs of the machine, of the technology, have tended to take precedence over the needs of us, the people.

Consider the typical manner in which the technology of automation is applied. In factories, in the control of ships and aircraft, in bank processing centers and insurance agencies, machines are used wherever possible to handle the large volume of information that passes through the offices, to control the assembly of parts, and to control the piloting of ships and aircraft. But technology cannot do the entire set of tasks. Even the smartest of computer systems is relatively single-minded, incapable of self-direction, learning, or flexibility. Everything has to be given to it "just so," or else it won't function.

In aviation, airplanes are flown more and more by automated controls. Did the designers do a careful analysis of the tasks faced by the pilot and decide which were best done by people, which were in need of some machine assistance? Of course not. Instead, as usual, the parts that could be automated were, the leftovers were given to the humans.

Worse, the automation works best when conditions are normal. When conditions become difficult—say, there is a storm and an engine, a radio, an electrical generator fails—then the automation is also likely to fail. In other words, the automation takes over when it is least needed, gives up when it is most needed. When the machine fails, often with no advance warning, people are suddenly thrust into the process, suddenly asked to figure out the current state of the system, what has gone wrong, and what should be done.

When people work on the factory floor, they can tell what is happening by sounds, vibrations, even by smells. In many computer-controlled factories, the computers are located in air-conditioned, air-filtered rooms, away from the heat, noise, and vibration of the factory. The people who used to work on the factory floor moved, to be with the automation, thus changing their interaction with the machinery. Whereas before they were physically able to keep an eye on things, often catching potential problems before they arose, now they were disconnected from the real world, watching second- or third-order representations: graphs, numbers, trend lines, flashing lights.

There are mixed virtues to dealing with factory processes in this manner. It does make normal operations proceed more smoothly. Some second- or third-order representations are able to present information and relationships that could not be appreciated or even discovered before in visible, useful ways. Alternatively, some representations impoverish their users, providing insufficient information to understand the entire problem, reducing the richness of sensory information, isolating the users from the situation.

One of the principles of human-centered design is that the visible, surface representations should conform to the forms that people find comfortable: names, text, drawings, meaningful naturalistic sounds, and perceptually based representations. The problem is that it is easiest to present people with the same representations used by the machines: numbers. This is not the way it ought to be. Sure, let the machines use numbers internally, but present the human operators with information in the format most appropriate to their needs and to the task they must perform. Each part of the system, human and machine, should be able to use whatever representation it finds most efficient, but it is the machine (and its human designers) that should take the extra steps to do the translations from the machine-centered form internally to the human-centered form at the surface.

Sushana Zuboff's influential book *In the Age of the Smart Machine* distinguishes between systems that automate and that informate. The word *informate* was invented by Zuboff to describe the potential of new technologies to inform, to provide people with rich access to a variety of information that would not be available without the technology. An informating system is a reflective cognitive artifact that gives people access to the knowledge they need to make informed, intelligent decisions. With an informating system, people can go beyond the requirements of a job to understand the larger picture, sometimes thereby improving and enhancing the work they do. Systems that informate tend to be cooperative ones, with person and system working together. Automated systems are not cooperative. They tend to be autonomous. They are not designed to assist the human, they are designed to replace the human.

It will take extra effort to design systems that complement human processing needs. It will not always be easy, but it can be done. If people insisted, it would be done. But people don't insist: Somehow, we have learned to accept the machine-dominated world. If a system is to accommodate human needs, it has to be designed by people who are sensitive to and understand human needs. I would have hoped such a statement was an unnecessary truism. Alas, it is not.

THE LANGUAGE OF LOGIC

There is a children's puzzle in the United States that goes like this:

> I have two coins that total 30 cents. One of them is not a 5-cent piece. What are they?[1]

Most people can't solve the puzzle. Obviously, there is a trick, but where? What could the answer be?

The answer is that one coin is a 25-cent piece, the other a 5-cent piece. "But," people protest, "you said there wasn't a 5-cent piece!" Not so. I said, "One of them is not a 5-cent piece," and that's true—one isn't a 5-cent piece, it's a 25-cent piece. It's the other one that's a 5-cent piece.

Misleading? Yes. Unfair? Yes. Proper and legal according to the laws of logic? Yes.

[1]For those not familiar with American coins, the relevant ones are: 1 cent (the penny), 5 cents (the nickel), 10 cents (the dime), 25 cents (the quarter).

From the seventeenth-century views of Descartes through today, the human mind has been thought of as a computational device, usually a rigid, computational mechanism based on clockwork or simple logic. Almost every advance in the science and technology of computation, control, and communication has also been described as an advance in the science of thought processes, usually without any evidence, usually by people who had never studied people.

These advances in knowledge have been substantial, and they certainly add to the general understanding of how intelligent devices could operate, how computations can be performed, and the basic properties of communication systems. These have provided valuable tools for the study of thought processes—whether real or artificial, human, animal, or machine—but that does not mean that they are models of those processes. The thought processes of humans are not like the mathematical logic of machines. Indeed, were the thought processes of humans like that of logic, we wouldn't have needed to invent logic as an aid to thought: Logic is important *because* it is different.

Logic is most definitely not a good model of human cognition. Humans take into account both the content and the context of the problem, whereas the strength of logic and formal symbolic representation is that the content and context are irrelevant. Taking content into account means interpreting the problem in concrete terms, mapping it back onto the known world of real actions and interactions. The point is not simply that people make internal mental models, stories, or scenarios of the problems they are attempting to solve but that they make special kinds of models: People map problems back onto their own personal knowledge and experiences.

I examined the complementary nature of logical analysis and stories in Chapter 5. The point is that each plays an important, but different, role in human thought. Logic deliberately abstracts the critical, quantitative aspects of the situation, providing a general method of reaching a conclusion that is independent of subjective biases and opinions. Logic is reliable: Provide the same information, and it will always reach the same conclusion. Stories, on the other hand, can emphasize the special aspects of the situation, the critical details that characterize the human side of the matter. Stories emphasize the qualitative aspects of the situation, the subjective biases and emotions of the participants. Stories are not reliable in the same manner as logic. The same situation will often elicit different stories from the people involved. The same stories will sometimes lead to different conclusions, depending upon the mood and character of the audience. The problem with logic is that it is too fixed and rigid, and any information that cannot fit into its rigid framework is thereby excluded from playing a role in the conclusion. The problem with stories is that they are too flexible, too subjective: Almost any point of view or conclusion can be buttressed by an appropriate story. In the end, we often need both—the hard, formal process of logic and the soft, subjective impressions from everyday experience.

Human language is also very different from the language of logic. Human language takes into account the point of the encounter, which is to communicate. Basically, the idea is that most of the time, we treat the person we are talking with as an intelligent partner who shares considerable knowledge about the topic and who understands the nature of communication. Both parties actively interpret the other's actions and utterances, the result being a shared space of ideas and concepts. We do more than work out the meaning of the words: We also try to figure out why they were said. If others tell us something, we assume it's for a reason. In fact, when others do not tell us something, especially if we know that they know it, the omission is also meaningful. Just what kind of information is conveyed follows some simple conventions, which can be summarized like this:

- People tell others what they think needs to be known. Not too much more—except to give the context and the reasons. Not too much less—except to spare the listener from items they believe don't need to be known (or that they do not wish to tell).
- No games. Everything is to be believed and is to be interpreted through everyday language.

- If there is some special condition, it will be described. If special conditions aren't mentioned, then the listener can assume there aren't any.
- If things are obvious, unimportant, or unlikely, they won't normally be mentioned. So when people do mention them, it conveys a special signal about their importance.

If I tell you that I have two coins, and one of them isn't a 5-cent piece, in normal conversation I mean that neither one is a 5-cent piece. I would never mean that one was and one wasn't: what a weird thing to mean. After all, if I had meant that one was a 5-cent piece, I would have said so: "I have two coins that total 30 cents. One of them is a 5-cent piece. What's the other one?" Logic, however, takes everything seriously and makes no assumptions about what could possibly be meant. You say it and you mean it. Period. "After all," says the logician, "we should communicate in clear and unambiguous terms. A word ought to mean the same exact thing each time it is used."

Nonsense. If we followed those rules, language would suffer in grace and flexibility, robustness and beauty. Language can be short and simple, simplifying the task of conveying ideas, and it can also be robust, insensitive to errors in speech, words, or grammar; insensitive to noise that obscures the sounds or distractions that cause the listeners' attention to wander. People tend to minimize mental work. George Kingsley Zipf called this "the law of least effort." Zipf pointed out that the words that occur most frequently in language are also the shortest. Words that start out long—such as *automobile* or *television*—become shorter as they become more popular: *Automobile* becomes *auto* or *car; television* becomes *telly* or *TV.* Pronouns are excellent examples of how we substitute short words—a pronoun like *it*—for an item or a concept that might take several words to convey. Some languages even do without the pronoun. In American Sign Language, if one is talking about two people, each person is given a location in space, and once a person has been talked about—signed about—further reference to that person is made simply by pointing to or moving the hands so as to do the signing at the appropriate place. These shorthand conventions add ambiguity to the language, which makes its formal, scientific analysis very difficult. In fact, we do sometimes have problems understanding one another. We sometimes misinterpret or ask for clarification. But these occasional problems are minor, and the natural processes of language manage most minor difficulties so smoothly that the people involved often do not even notice the problems. The simplifications are worthwhile.

The language of logic does not follow the logic of language. Logic is a machine-centered system in which every term has a precise interpretation, every operation is well defined. The operations are defined to emphasize consistency and rigor, so that no contradictions are possible, no ambiguities. Everything stated in logic has an exact and unambiguous interpretation, and its truth value, whether it represents a true or false statement about the world, can be precisely computed. Actually, there is more than one form of mathematical logic, but they all share the properties of mathematical precision and rigor. Even those forms of logic specifically designed to handle the ambiguities of language, such as "fuzzy" logic, do so in very precise ways, with precise rules by which the "fuzziness" of set membership and set operations is computed. Logic is very intolerant of error: A single error in statement or operation can render the results uninterpretable.

Language is quite different. Language is a human-centered system that has taken tens of thousands of years to evolve to its current forms, which exist in the multitude of specific languages across the globe. Language has to serve human needs, which means it must allow for ambiguity and imprecision when they are beneficial, be robust in the face of noise and difficulties, and somehow bridge the tradeoff between ease of use (which argues for short words and utterances) and precision and accuracy (which argue for longer, more specific utterances). Ease of use tends to win.

Language has other interesting constraints. First, it must be learnable by young children without formal instruction. This imposes as yet ill-understood constraints on the language structure. Second, language has to be malleable, continually able to change and adapt itself to new situations. Whenever two language groups come in contact, each language borrows from and adapts to the other. New technologies and new experiences require new words and perhaps even new grammatical structures. Third, human language has to be very tolerant of error. People often use the wrong words, use inappropriate grammar, change their minds halfway through an utterance and restart, yet all with minimal effect upon the listener or upon the accuracy of the communication. Finally, language has to fit the cultural context within which it is spoken. Some languages have special forms of honorifics that reflect the social categories of the culture. Other languages provide alternative structures that match their cultural practices. Many statements are intentionally ambiguous. Thus, many disagreements are eased by stating the resolution in deliberately ambiguous language that allows each party to claim satisfaction. The ambiguity fools nobody (except, perhaps, the onlooker), but it smoothes the social interactions. This tactic is widely used by legislative bodies. (The ambiguities are then left for law courts to interpret.) Language is a continually evolving, ill-defined, amorphous system, exquisitely suited for human interaction.

SOFT VERSUS HARD TECHNOLOGY

Technological systems can be classified into two categories: hard and soft. Hard technology refers to those systems that put technology first, with inflexible, hard, rigid requirements for the human. Soft technology refers to compliant, yielding systems that informate, that provide a richer set of information and options than would otherwise be available, and most important of all, that acknowledge the initiative and flexibility of the person.

When I speak of the power and virtues of technology, I am referring to soft technology: technology that is flexible, that is under our control. Hard technology remains unheedful of the real needs and desires of the users. It is a technology that, rather than conforming to our needs, forces us to conform to its needs. Hard technology makes us subservient; soft technology puts us in charge. Automating tends to be a hard use of technology. Informating tends to be soft.

There are problems with any technology, hard or soft, problems that arise from the general conflicts and differences in needs within society. What is best for one group may not be best for another. Necessary tradeoffs abound and often lead to difficulties. Hard technology is often superior for the person or group that has requested it, inferior for those who must then obey it. Sometimes hard technology is the correct choice. Accounting and timekeeping rules are hard technology, imposed for the benefit of an organization, often detrimental to those who must obey its dictates. Traffic lights and signs are hard technology, inconvenient at times for the individual who must stop at a deserted street intersection some lonely 3:00 AM, but beneficial for society.

As I discussed in Chapter 8, privacy is a cultural convention, one that differs across societies. Some societies have essentially no notion of privacy, others go to extremes. In my culture, we expect considerable privacy. When people telephone me at home to ask a business question, they usually first apologize for the intrusion. The apology reflects the cultural bias that separates work and family and assigns a value to privacy in the family. I expect to be able to work at home in private, unwatched or unnoticed by others. At my office, I do not expect the same privacy. Even so, before I enter someone else's office, I request permission. When I see someone else engaged in conversation, I do not interrupt, or if I must, I apologize.

The Telephone System

The technological affordances of the telephone do not allow the caller to follow the normal rules of social courtesy. When the caller makes the call, there is no way of knowing the activities of the recipient, no way to tell whether the person is busy, engaged in conversation, unwilling to be interrupted, or anxiously awaiting the call. Until recently, the recipient had no way of knowing who was calling or for what reason. The shrill sound of the bell simply informs that someone is on the line, someone very important or critical, or someone trivial or self-serving—a salesperson, perhaps. Or even a wrong number. Conversations get interrupted. Thoughts get disturbed, tranquillity violated. This is convenient for the caller, but inconvenient for the recipient, and least convenient of all for those trying to interact with the person being interrupted. The result is that the telephone has destroyed many common courtesies. Yes, the caller can apologize and ask permission to continue the call, but the interruption has already taken place. The traditional telephone is a hard technology.

Is it possible to develop a softer, more humane technology for the telephone? Yes. Today the telephone is designed around the needs of the telephone system: The goal is to make proper connections as efficiently as possible in order to minimize the demands on and cost of the equipment at the central telephone switching offices. Whether this serves the needs of either participant is not considered, but if anything, it favors the calling party, who after all, has control over when the call shall be initiated.

Actually, the caller does not always wish to contact the recipient. At times, it is more efficient to leave a message without having to converse or without having to explain to various intermediaries who might answer the call. Wouldn't it be nice if the caller could specify whether the call is intended for a message center (an answering service or machine or voice mail) or for a particular recipient? Similarly, from the recipient's point of view, wouldn't it be nice to know who was on the phone, and maybe even why, before deciding whether or not to answer? Some of us already accomplish this by letting all calls go to an answering machine. The answering machine plays aloud any message the caller leaves, affording call screening: The recipient can privately listen to the message, picking up the telephone and talking with the caller if the call is important or of interest or relevance, postponing the remaining calls to a more convenient call-back time.

Now that many telephone switching offices are converting to digital technology, it is possible for the telephone number of the calling party to be displayed to the recipient. As with all new aspects of a technology, this service has both virtues and drawbacks. It is an obvious convenience for the recipient of the call, but it poses loss of privacy for the caller. When one private person calls another, privacy is not an issue. In fact, it is not polite to disguise one's identity. But what about when people call businesses or government agencies and health clinics? When people call a crisis help line, shouldn't they be able to keep their telephone number secret? The number of people willing to use these social services might diminish if they could not call anonymously. And what about calls to business or government? I do not want a simple request for information to result in my name being added to yet another computer-maintained list that thereby increases the general information about my habits and leads to yet more advertisements and sales calls arriving at my home.

If you consider the issues, it becomes clear that identifying the telephone number of the caller is the machine-centered approach. This is the information that the system finds easy to provide. But is this what the call recipient wants to know? Not really: We want to know who is on the line, not the number that identifies a particular telephone. Of what good is the phone number? Now we are forced to learn yet more numbers (or program them into our telephone) so that each number could be translated into a person-centered identification of who might possibly be at that number. The phone number is really of use only to businesses or law-enforcement agencies. In fact, friends might call from a variety of places and numbers—how could we keep track of all the possibilities? No, we only want to know who is calling.

Would identifying the name of the caller still present a privacy problem? Yes, but one that is a lot easier to deal with. The name alone does not uniquely specify how to reach someone, so it would not be useful for most data records. And when extreme privacy was desired, people could use nicknames or pseudonyms, names that would be identifiable to their friends and colleagues, but not to others.* This scheme protects callers and recipients alike: Callers provide only as much information as they are comfortable with; the recipients can then decide whether or not to accept such calls. There is no advantage to callers in lying about their identity, for the proper remedy in such a case is for the recipient to hang up immediately.

How can the identity of the caller be provided? Several schemes are possible. Here is one: Suppose that upon placing a call, the caller could also give a three-second message that would get delivered along with the ring. The recipient would hear "*ring* . . . This is Julie for Don . . . *ring*" or perhaps "*ring* . . . Petersen's Auto Repair with a question about your car . . . *ring*." Better yet, just as in normal conversation, suppose the ring were really interpreted as a request to schedule a conversation, not, as now, a demand for instant talk. Let the recipient choose whether to answer or to signal something like "In five minutes, please." The caller could decide whether to call back then (the telephone could do that automatically) or to deliver a message to the message center. This is a soft, person-centered technology: Caller and recipient alike could use it in whatever way best fit their needs.

As with all new technologies, the suggestion that a caller be allowed to insert a brief voice message into the ring sequence poses some problems. It has interesting privacy ramifications at the receiving end if unintended people hear the message. It would allow brief long-distance messages to be delivered without an official phone call taking place. It is subject to misuse. Just as cognitive artifacts change the task to be performed, the communication technologies change the nature of human interaction. Both society and technology mutually adapt to their presence.

There are many other possible aids to telephone communication, some of which are starting to appear in today's systems. The technical capabilities of the phone are presently limited by the low communication bandwidth of the wires that connect the home to the central office. The telephone signal is of rather poor quality: The voice is tinny, and any other signals—such as computer, facsimile, or television signals—can only be sent slowly, usually with reduced quality. In addition, the telephone instrument itself has very little capability to communicate with the person who is using it: The human-sided, soft capabilities are limited by the inferior capabilities for communication between the telephone instrument and the user. Soft technologies often require more powerful technologies than do hard ones: In the case of telephones, the sets have to provide more information about the calling party, more controls, and better information displays. The limited keypad and lack of visual displays of the standard home telephone make it almost impossible to humanize the telephone. In general, in softening a technology, the most critical thing is to start with the needs of the users, then employ those needs to determine the technology that will be provided.

The hard technology of the current telephone system gives us little choice but to violate privacy and standard cultural behavior. The soft technology of the future telephone has the potential to put people back in control.

APPROPRIATE TECHNOLOGY

Is there a way to transform the hard technology of computers and information processing into a soft technology suitable for people? Yes, I think so. The correct approach, as I just argued in my discussion of the telephone system, is to start with the needs of the human users of the system, not with the requirements of the technology. With some thought, it is possible to transform even the most inhumane of systems into ones that are quite acceptable. Consider the post office stamp machine.

The Stamp Machine

In my book *Turn Signals Are the Facial Expressions of Automobiles* I told the story of the United States Postal Service and its automated vending machines for stamps. The story is appropriate for inclusion here. In brief form, this is what happened. The Postal Service installed machines so that patrons could purchase stamps when the post office was closed or, for that matter, avoid long lines even when the post office was open. But soon after the machines were installed, I started hearing stories about them on one of the computer newsletters to which I subscribe.* Disaster at the stamp machine.

Filled with curiosity, I went over to my local post office and looked at the machine. Hmm. There certainly was a problem. Rule of thumb for design: If people have pasted signs on a machine, there is something wrong with the design. The machine at my post office in Del Mar, California, had not only hand-lettered signs on it but a fancy, computer-controlled sign with scrolling red letters that said:

> WELCOME TO THE DEL MAR POST OFFICE VENDING MACHINE *** I REFUND A MAXIMUM OF $3.25 CHANGE WITH YOUR PURCHASE *** THINK BEFORE DEPOSITING A BILL LARGER THAN $5 ***

Now put yourself in the place of a postal patron who has just inserted $30.00 into the machine in order to purchase a roll of one hundred 29-cent stamps, expecting to get the stamps and $1.00 in change. But then, after the machine has graciously accepted the money, it informs you that it no longer has any of those rolls: What would you like to buy instead? And, no, it can't simply return your $30.00 (it returns no more than $3.25, remember?).

The post office stamp machine is clearly a machine-centered design. The designers hadn't even learned one of the most fundamental lessons of human-centered design common in the computer industry: reversible operations. Today most decent computer programs and operating systems come with "undo" commands. Undo, as in, "Oops, I didn't mean to do that—undo it, quick." We have reached the point in computer systems where users expect undo commands: Reviews in computer magazines of programs that lack the undo feature often advise their readers not to purchase them. Alas, other industries have not learned from the experience of the computer industry. The vending-machine industry, in particular, seems not to care. Once it has your money, it never wants to let go of it.

How could the post office machine be made better, softer? There are several answers, but the real point is, once again, that the machine, any machine, should be designed by starting with the needs of the people. This is yet another machine whose designers narrowly focused only upon the machine's needs. The machine needs money and the users' choices, so the users are supposed to provide them: Put in the money, say what is wanted, and no back talk. Notice that the machines require the money first, then the selection of what you want, in contrast to people-to-people operations.

A human-centered, soft design would let users decide in what order to do things: money first or selection first, whichever was wanted. User control, not machine control. If users made the selection of items first, they could get immediate confirmation of whether the items were in stock. In either case, whether money first or selection first, the users should always be allowed to change their minds: There should be two buttons—one marked "PROCEED WITH SALE," one marked "CANCEL SALE." If the sale is canceled at any time prior to delivery of the items, all the money should be returned. These simple changes would already soften the machine. They aren't sufficient, for a few hours of observation of the stamp machine in operation reveals numerous other problems, but all could be solved—or at least softened. In many cases, it isn't hard to transform a machine-centered design into a human-centered one: It just takes some thought, care, and understanding of the needs of the people the machine is intended to serve.

There are many positive examples of appropriate technology. Many excellent examples come from information technologies that provide us with rich information about a topic of interest, leaving us in control of how to perform the search and what to do with the results. The daily newspaper is perhaps one excellent example of a current database that allows ready access to the latest news, weather, radio and television schedules, and information about events of the world and of the local region. The technology is non-obtrusive, perhaps because it has no choice: It is a passive, surface artifact, so that all the major work has to be done by the user. This does not stop it from being very effective.

I have already discussed how dictionaries and reference books might be enhanced by being made available electronically, with appropriate "agents" that would allow users to explore the contents at will. The trick is to provide tools appropriate to the way people normally think. Instead, most computer systems tend to be designed with the efficiency of the system in mind. Considerable training is often required before they can be used, primarily because they force us to formulate questions using the unnatural and awkward precision of logic. Most library retrieval systems and database query languages do not meet my criteria for human-centered design. One problem with these systems is that they insist upon more precision and specificity than we may wish to provide. We often are not really certain of either the question or the answer: That is why we are looking. In a proper system, the process of exploration will let us discover the question as well as the answer.

Rabbit

One of my long-term favorite examples of a system that attempts to provide a soft, appropriate technology is illustrated by a computer software system developed by Michael Williams and Frederich Tou, then at the Xerox Palo Alto Research Center. This system, which the authors named Rabbit, attempts to provide information that you need to make a decision. Unlike other such services, it does not assume that you know anything at all about the way it stores its information. Instead, you simply request whatever you think is relevant and the system will respond with samples.

Suppose you were hungry, in a strange town, and wanted a restaurant. You came across Rabbit. In most systems, you would have to provide a precise query, and so you would first have to learn how the system was organized. Not so with Rabbit. You could start off simply by saying, "I would like to go to a restaurant." This is really an inadequate request, because it does not specify the location, type of food, and other variables based on which the system expects to select. Worse, there might be a thousand restaurants in the database. How would the system know what to do? Most systems would either reject the request or provide all thousand entries. Neither is very satisfactory.

Rabbit, however, reverses the normal interaction. Rabbit is not trying to choose for you, but rather to help you choose for yourself. So if you simply ask for a restaurant, Rabbit will give you one, any one. It doesn't even matter what restaurant it gives you, because it is teaching you about the kinds of information it needs. Suppose it responds to your general query with a very formal Chinese restaurant located on the other side of the city that requires advance reservations:

Name:	Chan Wo
Location:	1800 W. Far Side Drive
Price:	$$$$ (very expensive)
Formality:	Suit required
Type of food:	Chinese

Service:	Superb
Reservations:	Highly recommended
Credit cards:	No

You may still not know what you want, but one thing you do know is that this is not it. Good! Rabbit works this way deliberately, for it can learn your desires by taking account of the things you don't like. Either you can simply reject this choice and ask for another, or better yet, you can use the restaurant description as a starting point by indicating just what features of this selection you like, dislike, or are indifferent to. The way this works is that you can select any of the descriptive dimensions and modify it. Rabbit helps by giving a set of alternatives: For *Location*, it can list the areas of the city and let you say for each area whether it is acceptable, not acceptable, or doesn't matter. For *Type of food*, once you saw the list of possibilities, you could rule out Chinese while accepting Thai, rule out French while accepting Northern Italian, and say "Don't care" regarding the rest.

Location:	I want the East Side.
Price:	Must be inexpensive ($).
Formality:	I prefer informal.
Type of food:	Not Chinese. Thai is OK. Not French. Northern Italian is OK. Otherwise, I don't care.
Service:	Don't care.
Reservations:	Must not be required.
Credit cards:	Must be allowed.

Rabbit is educating you, without your realizing it, but on your own terms, in your own language. The suggested alternatives are structured to teach you about the way Rabbit organizes them, allowing you to adjust and use its organizational structure to help refine your search. "Oh," you might think, "credit card. I forgot about that." You then tell Rabbit, "I don't have much cash, so only show me places that accept credit cards." Notice too that the system doesn't ever have to get to a unique answer. At some point, you could simply ask for several examples, read them, and decide among them, never telling Rabbit your decision.

Although Rabbit never asks its user to present a request using the language of logic, Rabbit is internally constructing its own logical expressions to describe your likes and dislikes. That is fine, for logic offers an important precision and value to the computer system. Logic is not the appropriate tool for most people, but it is quite appropriate for machines.

Rabbit provides a positive example of a person-centered, appropriate technology. Give control to us as individuals, in our language, and on our terms. The technology does the translation from its internal machine-centered logic to a form appropriate for the person. Rabbit was only used in the research laboratory as a demonstration: It has never been used in a commercial system. Too bad.

TEN

TECHNOLOGY IS NOT NEUTRAL

TECHNOLOGY IS NOT NEUTRAL. EACH TECHNOLOGY HAS PROPERTIES—affordances—that make it easier to do some activities, harder to do others: The easier ones get done, the harder ones neglected. Each has constraints, preconditions, and side effects that impose requirements and changes on the things with which it interacts, be they other technology, people, or human society at large. Finally, each technology poses a mind-set, a way of thinking about it and the activities to which it is relevant, a mind-set that soon pervades those touched by it, often unwittingly, often unwillingly. The more successful and widespread the technology, the greater its impact upon the thought patterns of those who use it, and consequently, the greater its impact upon all of society. Technology is not neutral; it dominates.

IS THE MEDIUM THE MESSAGE?

Marshall McLuhan proclaimed that the way in which we interacted with the events of the world was as important as, or perhaps even more important than, the events themselves. "The medium is the message," he proclaimed in a famous, often-repeated sentence.

I don't agree. In fact, I was tempted to title this chapter "The Medium Is Not the Message." I prefer to believe that the message is what you make of it: The medium is the carrier, not the contents.

Nonetheless, the medium is not a neutral carrier; it has numerous properties that affect both how it is used and its impact upon society. There is much to be said for McLuhan's argument that the medium changes how we interpret the message. In fact, many of the arguments in this book support just this point. Media can deceive and seduce even the discerning mind.

Each technological medium has affordances, properties that make it easier to do some things than others. McLuhan's famous remark to a large extent was a comment on affordances. Let me illustrate the role of media affordances by contrasting reading the printed page with viewing a television show. Each medium, the printed page and television, has both a positive side and a negative side.

On the positive side, reading affords control of pace: You, the reader, control which portion of text is read, which portions are skipped, which repeated. At any moment you can stop your reading and contemplate what has just been read. You can question, ponder, disagree. Reading affords reflection.

On the negative side, reading is relatively slow, relatively difficult. It takes considerable training and practice to learn to read. It takes mental effort, even for the most skilled reader: When given something to read, haven't you ever said, "I'm too tired to read right now"? The mental demands of reading preclude other simultaneous activities involving thought or vision. Reading requires concentration, a focus of attention upon the material. Written material tends to be information-rich, so that considerable mental activity is needed to decode the author's message. Obviously, some texts are more complex than others—for instance, look at the difference between comic books and textbooks—but on the whole, reading affords mental concentration and effort.

Printed text has a number of limitations as a tool for reflective thought. It is only a display medium: The words are fixed, unchangeable. Sure, you can cover them up or expand upon them through written comments in the margin, but they cannot respond. Remember what Socrates, the great Greek philosopher, said about written words (see Chapter 3): "[I]f you ask them anything about what they say, from a desire to be instructed, they go on telling you just the same thing forever." This does not mean that you have to accept everything you read unquestioningly, but it does make sustained debate difficult, if not impossible. In this sense, reading does not afford the same kind of prolonged, reflective debate and argument possible with an interactive medium, such as another person.

Now consider television, especially broadcast television. This is also a display medium, but one with very different properties from the printed text. Where reading is relatively difficult, television is relatively easy. You had to be taught to read, but no training or practice is required to watch television: Just plunk yourself down in front of the image and look. It does not take much mental effort. In contrast to reading, which you might not feel like doing when mentally fatigued, you might very well turn to the television set "for relaxation." As for pace, the two media are dramatically different. Where reading is self-paced, under the control of the reader, television is event-paced. The viewer has no control: The material flows relentlessly, continually driving the senses. There is no time for reflection, no time to ponder or reconsider. You cannot stop the flow, at least not without missing something else. As Jerry Mander put it in his critique of technology: "The information on the TV screen—the images—come at their own speed, outside of the viewer's control; an image *stream*. One doesn't 'pull out' and contemplate TV images . . . the nature of the experience makes you passive to its process, in body and mind."

Mander is describing the worst excess of an experiential medium: pure accretion of knowledge without time for review, reflection, or even assessment. According to Mander, the end result of years of television watching is a major impact upon our knowledge and understanding of the world, and from his point of view, the impact is almost entirely negative.

Mander continues: "The environment of TV is not static, it is aggressive. It enters people's minds and leaves images within, which people then carry permanently. So television is an external environment that becomes an internal, mental environment." All of Mander's arguments about the medium of television are the same as my arguments about the distinction between experiential and reflective thought. Nonetheless, I don't completely agree with his conclusions. Mander focuses upon the extreme, upon the excesses of shallow material presented so as to exploit the seductive powers of the experiential mode. But there is a positive side to television as well as a negative one: Television, properly constructed, can be a powerful tool for reflection. Let me explain.

Reflective Thought

The human mind, for all its powers, is limited in its ability to think deeply about a topic, primarily because of the restricted capacity of working memory. Try to hold something in consciousness, and as soon as you are distracted by something else, it fades away. Try to hold onto too many thoughts at once, and they interfere with each other. The psychological evidence is that only about five things can be kept active in consciousness at a time. Five.

One method for expanding the power of the unaided mind is to provide external aids, especially notational systems, ways of representing an idea in some external medium so it can be maintained externally, free from the limits of working memory. The printed words you are now reading are one example: They represent the mental structures in my mind that I am trying to transfer to yours. The power of external representations is that you do not have to keep the material in mind—the piece of paper on which it is written keeps it for you.

Writing has the virtue that the ideas are permanently marked on paper. If I read a story or an essay or a scientific report, I can review it, compare this section with that, analyze the structure and the content. This would be difficult or impossible if all we had available to us was the unaided mind. The external representation makes all the difference.

Not all intelligent systems are capable of reflection. Most animals cannot reflect or have only limited abilities, at best. Most computer programs cannot. For a system to be capable of reflection, it must satisfy some technical requirements. It has to have an internal representation of knowledge and the ability to examine, modify, and compare its representations. In technical language, the system must have a "compositional" representational medium that allows (affords) adding new representations, modifying and manipulating old ones, and then performing comparisons. The human mind is a compositional medium.

Paper and pencil can augment human reflection because they too constitute a compositional medium. Television cannot, at least not the normal home television, which does not afford composition and therefore does not afford reflection. Notice, by the way, that the printed book, by itself, is not a compositional medium, because the words are fixed and not easily changed. The paper on which the words are printed does afford compositional acts, but only when we take pen or pencil to it.

Reflection requires more than a compositional medium: It requires the time and ability to elaborate upon and compare ideas. The medium must afford the time for reflection. Here is yet another contrast between books and television. Reading affords reflection by coupling the self-paced nature of the act with the compositional powers of the mind. Normal television is a display-only, event-driven medium, well suited for experiential cognition, but not for reflection. The medium is in control of the pace, so although, in principle, the television image could be coupled with the compositional powers of the mind for reflective thought, the event-driven pacing does not afford the time required for reflection.

Interactive, viewer-paced television can afford reflection if the viewer has essentially the same control over the images as does the reader over which part of the text to read. Interactive television does have this power to allow the viewer to select what is to be seen and to control the pace of the material, making it just as easy to skip back and forth as with the print medium. Consider the story of "The Shakespeare Project."*

One of the strengths of serious literature is that alternative interpretations are possible. The reader's understanding of the characters and of the societal issues being addressed by the work is enhanced by exploring alternatives to the possibilities suggested by the author. This requires time to stop and reflect upon the issues, to question and explore. This is difficult to do when watching a single performance of a play, movie, or television show, but it is made possible when alternative performances can actually be viewed and contrasted. A reading or viewing of a play provides an interpretation of the events and characters. The problem is, this interpretation then traps its audience into holding this single point of view without consideration of alternative possibilities. A good instructor tries to break students out of a single interpretation, but students often resist: The interpretation they have is so compelling it is difficult to understand how others could exist. Seeing the play is worse than reading it in this regard because the actors have presented a well-worked-out instance of a single interpretation.

But suppose you (or the student or instructor) have available recordings of many performances coupled with the appropriate technology to allow you to read the written version, then observe several performances, each with a different interpretation. Suppose you could go to any section of the written play and immediately call up alternative performances, allowing you to compare how different acting groups interpreted the parts. Now you have a tool for reflection, allowing you to compare and contrast different interpretations in a manner superior to that possible with the written word alone. Aha! Now television is a visual technology that can add to understanding, that affords reflection.

The suggestion that television might be a medium for reflective thought is not idle speculation. My description comes from an instructional system for teaching theater called The Shakespeare Project, developed by Larry Friedlander at Stanford University. Friedlander combined several filmed theatrical performances onto a videodisc, all controlled by a computer that can also display the script and allow selection of the scenes to be watched. In my first interaction with the program, I learned things about *Hamlet* that none of my prior readings and viewings had suggested to me.

The difference between this use of television and our normal, everyday use is that The Shakespeare Project provides an interactive, reflective use of television. It is self-paced, thereby allowing time for reflection. When I tried the system, I could stop and start the performances at will, switching among them in whatever manner was most appropriate to the questions I generated about the play. I could compare different points of view as slowly or rapidly as I wanted. I could dictate my own pace. One section of the program asked me to use the computer keyboard to write my interpretations of the thoughts of each character at each moment throughout the play, then compare my impressions with those of directors and actors. In this case, the medium of television was combined with reading and writing, capturing the best of both worlds.

Because the human mind is a reflective, compositional medium, it can transform any technology into a reflective one—if the technology permits sufficient control of pace to allow the time required for reflection. Although reflection can be done with the unaided mind, the power is greatly enhanced by external representations provided through technology—but only if the technology is appropriate. All technologies have their equivalents of the comic book or soap opera. The difference between the experiential nature of some books or television shows and the reflective nature of other books and interactive television is partially in the technology, partially in the mind of the person.

APPROPRIATE USE OF TECHNOLOGY

Although I may seem to view much of technology with skepticism and despair, overall I am an optimistic proponent. The various technologies of cognition can provide us with useful powers for acquiring, using, and creating information and knowledge. If employed properly, they can enhance the quality of life. If employed inappropriately, they diminish it.

But in saying this, am I not succumbing to the "everything is OK" philosophy? Mander put it this way:

> I have attended dozens of conferences in the last ten years on the future of technology. At every one, . . . someone will address the assembly with something like this: "There are many problems with technology and we need to acknowledge them, but the problems are not rooted to the technologies themselves. They are caused by the way we have chosen to use them. We can do better. We must do better."

Moreover, after stabbing those of us trying to improve the use of technology, Mander twists the knife: "This is always said as if it were an original and profound idea, when actually everyone else is saying exactly the same thing."

Technology has made us smart, smart in the sense of being better able to think, to reason, to make judgments. But technology comes in many forms, and there is no question that it has negative consequences. Television can be an engaging diversion. Come home after a stressful day, turn on the set, and stay mesmerized for the rest of the evening. No thoughts, no worries. No more stress. This by itself might be regarded as reasonable therapy, until it becomes an addiction, carried out night after night after night. What goes on in the mind of the television viewer during all that time? "Guilt," say Kubey and Csikszentmihalyi, "guilt about TV viewing, especially excessive viewing, is fairly common among American, British, and Japanese respondents and is greatest for middle-class viewers."

The problem, as I see it, is not so much in television as in the experiential mode of behavior that it so nicely supports. The experiential mode is seductive; it draws viewers into its clutches, enticing them with pleasurable sensations, allowing time to pass by quickly, without effort; and the unceasing barrage of sensory information helps keep the mind from thinking about the concerns of life. In fact, that is why so many viewers feel guilty. "This guilt," say Kubey and Csikszentmihalyi, "is often tied to passivity, namely the feeling that one should have done something more productive than sitting or lying in front of the TV."

The experiential mode drives the entertainment factories of film and video: Entertainment is almost by definition experiential. Were it up to the entertainers, there would be hundreds of channels of full sensory barrages available to every human on earth, every second, every day. Full audio, full video, full motion, full experience: all the glories described in the chapters of this book, and then some.

Technologies are not neutral. They affect the course of society, aiding some actions, impeding others, independent of the morality or necessity of those actions. Technology also has its side effects, both physical and mental. Technology can aid as much as it can detract. It really is up to us, both as individuals and as a society, to decide which course we shall take. The course followed so far is not necessarily the appropriate one. Not all of the technology of cognition has failed. Some has served us well. There is an appropriate way to use technology.

The Human Side of Technology

Want a good example of a humane technology? Consider the handheld calculator. People tend not to be good at arithmetic, for all the reasons discussed in this book. Sure, you can multiply 5 times 47 in your head, but what about 279 times 725? Even if you use paper and pencil, long calculations are apt to contain errors. Arithmetic requires precision and accuracy, usually involving long strings of numbers that far exceed the capability of working memory. It is perfectly appropriate to turn over the tedium of long, complex calculations to machines: The precision, accuracy, and memory requirements are readily met by today's computational devices. Even small, inexpensive calculators far exceed the normal person's ability to do arithmetic. The nice thing about this technology is that it is unobtrusive, undemanding. We control when and how it is to be used, we control the pace. The calculator is an excellent example of a complementary technology, one that supports our abilities but does not get in the way.

Want other examples of appropriate technologies? The book. Writing in general. Or how about tools that have evolved to fit the true needs of the user, tools that have resulted from years of shaping by individual craftspeople? Examples include tools for carpentry and metalworking, tools for the farm and garden, for camping and mountain climbing, art and cooking. But to find the good tools, the ones as yet unblemished by fads or the need to put appearance before function, you must visit the stores for professionals: a professional restaurant supplies store (not the department store), a good old-fashioned hardware store, or a hiking and camping store. Here you can find tools that have slowly evolved over years to fit the needs of the people and the task, to be under human control.

Tools such as the calculator or book allow people to be in control. When they are needed, they perform their functions efficiently and smoothly. Otherwise, they stay quietly in their storage place.

Appropriate tools are designed by starting off with human needs, working with those who will be using the tools to fashion them into the most effective instruments for the task. Above all, such tools allow people to be in control: This is an appropriate use of an appropriate technology.

You might note that most of the tools that I cite as an appropriate use of technology are physical artifacts, not cognitive ones. In part, that is because cognitive tools, on the whole, are information-based tools, ones with interior representations and for which the design problems are complex. There is no folk design for cognitive artifacts as there is for so many hand tools, no equivalent of tools for the garden or for sports. Cognitive tools are simply harder to get right. Probably the closest equivalence is in the variety of specialized notebooks and daily organizers, oftentimes very well suited for a particular set of habits. The book and calculator have also benefited from years of slow modification: hundreds of years in the case of the book, tens of years for the calculator.

MAKING TECHNOLOGY HUMANE

The difficult problems are the social ones, not the technological ones. I am looking for a person-centered view of technology, not the machine-centered view that we now have. I don't want to see computers in the office, the home, the schools if they are there just for the sake of technology or, even worse, as a way of grabbing attention by glamour rather than by content.

Technology is used much too often for dramatic effect, without worrying about content: We let the entertainment industry drive the contents of our books, films, and now, computer media for schools. The entertainment industry understands experiential mode. The result: We are in danger of using the technology to amuse ourselves to death rather than to inform and enrich our lives.

In his book *Amusing Ourselves to Death*, Neil Postman reminds us of two contrasting views of the impact of technology on society: one described by George Orwell in his book *1984*, the other described by Aldous Huxley in *Brave New World*. Says Postman:

> Orwell warns that we will be overcome by an externally imposed oppression. But in Huxley's vision, no Big Brother is required to deprive people of their autonomy, maturity and history. As he saw it, people will come to love their oppression, to adore the technologies that undo their capacities to think. What Orwell feared were those who would ban books. What Huxley feared was that there would be no reason to ban a book, for there would be no one who wanted to read one. . . . In *1984*, Huxley added, people are controlled by inflicting pain. In *Brave New World*, they are controlled by inflicting pleasure.

I too fear with Postman the attractiveness of the experiential mode, the attractiveness of doing ourselves in with pleasure.

We have done more to ourselves than inflict pleasure. We have elevated mechanical, machine-centered modes of thought to the undeserved status of a model for people to emulate. Yes, we should take advantage of machines, of science and technology and the resulting formal tools for aiding thinking, planning, decision making, and design, but not to the exclusion of human values. When we take the machine-centered point of view, even if unwittingly, the result is a poor match with human abilities and needs. No wonder people make so many errors in their work, no wonder human error is considered the major cause of industrial accidents.

Remember the motto of the 1933 Chicago World's Fair (from Chapter 1): "Science Finds, Industry Applies, Man Conforms"? That was a machine-centered view of the world—unabashedly, proudly machine-centered. It is time to revolt. We can't conform. Moreover, we shouldn't have to. It is science and technology—and thereby, industry—that should do the conforming. The slogan of the 1930s has been with us long enough. Now, as we enter the twenty-first century, it is time for a person-centered motto, one that puts the emphasis right:

People Propose, Science Studies, Technology Conforms

CHAPTER NOTES

PREFACE

Page	Notes
xii	*This book is the fourth in a series:* Norman and Draper (1986), *User Centered System Design;* Norman (1990), *The Design of Everyday Things* (the hardcover title is *The Psychology of Everyday Things [1988]*); and Norman (1992), *Turn Signals Are the Facial Expressions of Automobiles.*

ONE

A HUMAN-CENTERED TECHNOLOGY

Page	Notes
3	*Twenty-one thousand commercials per year:* Mander (1991, p. 79).
8	*I am not the first to have pondered the duality of technology:* There are numerous critics, too numerous to mention. My favorite historical studies are Giedion's *Mechanization Takes Command* (1948) and the books of Lewis Mumford, especially *Technics and Civilization* (1934) and *The Myth of the Machine* (1967). More recently, we have Neil Postman (1985, 1992), most especially in his extremely important book *Amusing Ourselves to Death.* Along the way, we passed through Aldous Huxley (*Brave New World* and *Brave New World Revisited [1932, 1953]*) and George Orwell's *1984* (1950). The doom criers are also numerous, the most recent being Mander in *In the Absence of the Sacred* (1991) and Postman's *Technopoly* (1992).

9 *"Science Finds, Industry Applies, Man Conforms."* Cited in Pursell (1979). Also see Pursell (1990). I am indebted to Chris Spelius for bringing this to my attention.

13 *Around the start of the twentieth century:* Frederick Taylor was the most famous of the early time-and-motion study advocates in the United States. He was not alone, but he did more to popularize the American approach than any other person. In Europe, the advocates of "the science of work" claimed both precedence and superiority (for they did physiological recordings of fatigue and efficiency, whereas Taylor simply watched the motions). For the history and debate, as well as the implications for human behavior, see the engrossing book by Rabinbach (1990), *The Human Motor.*

The fascinating story of the introduction of the American capitalistic methods of "scientific work analysis" into the communistic Soviet Union, where it won the championship of Lenin, Trotsky, and Stalin—all three—is told in Chapter 6 of Hughes (1989, pp. 249–294): "Taylorismus + Fordismus = Amerikanismus."

16 *It is dangerous to divide something as complex as human cognition into only two categories:* The danger has not stopped many a thinker from doing so. The philosopher William James put it this way: "To say that all human thinking is essentially of two kinds—reasoning on the one hand, and narrative, descriptive, contemplative thinking on the other—is to say only what every reader's experience will corroborate" (quoted in Bruner, 1986, p. xiii).

In his book *Actual Minds, Possible Worlds,* Bruner (1986) argues that the two modes of thought are "paradigmatic" (scientific, logical) and "narrative" (stories, drama, experiences). Bruner's two modes are closely related to those of James and to mine, paradigmatic being the same as my reflective mode and narrative being related to, but not quite the same as, my experiential mode.

My distinction is also closely related to the distinction made within cognitive psychology between "controlled" and "automatic" processes. Brenda Laurel, in her provocative book *Computer as Theatre* (1991), talks of two modes of interacting with a computer: productive and experiential. Her experiential mode is limited to that of receiving the experience; I also allow activity on the part of the person. Thus my experiential mode combines both her experiential mode and aspects of her productive mode. For me, productivity can come about through either my experiential or reflective mode. Laurel's book provides an excellent introduction to the positive side of experiential interaction.

Jens Rasmussen, a Danish cognitive engineer, has long advocated the distinction among the three modes of operation: skill-based, rule-based, and knowledge-based (Rasmussen, 1983). His usage precedes mine and clearly influenced my own development. Rasmussen was concerned with the more practical problems of constructing appropriate control panels and equipment for use in industrial settings. Hence the difference in terminology, Rasmussen's in-between "rule-based" stage, and the different emphases.

TWO

EXPERIENCING THE WORLD

Page	Notes
21	*Something few other museums provide:* The new director of the museum, Goéry Delacôte, shares my distaste for the lack of reflective content of most museums and is instituting changes in the Exploratorium that are intended to remedy this problem. One museum that sounds as if it shares this philosophy is the Exploratory in Bristol, England, where Richard Gregory has worked to provide the proper spirit of exploration and learning aids: "hands-on science," he calls it. I haven't yet visited this museum, but it is high on the list of things to do next time I am in England.
21	*Note that my complaints are shared by others:* The quote in this paragraph comes from Gregory (1989). See the debate in the *New Scientist,* starting with the article in the October 5, 1991, issue by Paul Wymer entitled "Never Mind the Science, Feel the Experience" (and subtitled "Paul Wymer Feels the Time Has Come to Question the Purpose of Interactive Science Displays"). This was responded to in the October 26, 1991, issue by three letters (p. 65) and an essay by Melanie Quin, director of ECSITE, the European Consortium for Science, Industry and Technology Exhibitions (Richard Gregory, of the Bristol Exploratory, is the president of ECSITE). Quin's essay is called "All Grown Up and Ready to Play" (subtitled by the magazine "Melanie Quin Celebrates the Role of Interactive Sci-

ence Centres"). Quin lists a number of innovative science centers around the world: Singapore; San Francisco and Berkeley, California; Copenhagen; Toronto; Heureka—the Finnish Science Centre; Bristol, England; Perth.

21 *"Is the goal of motivation . . . reason enough for funding interactive science?":* Quin (1991, p. 61).

24 *Computer scientists Lokendra Shastri and Venkat Ajjanagadde have made an important first attempt:* See Shastri and Ajjanagadde (1993).

28 *Three different kinds of learning:* My colleague David Rumelhart and I explored the implications of this argument in two chapters: Rumelhart and Norman, "Accretion, Tuning and Restructuring: Three Modes of Learning" (1978) and Rumelhart and Norman, "Analogical Processes in Learning" (1981).

29 *It takes a minimum of five thousand hours to turn a novice into an expert:* Norman (1982).

31 *Experimental psychologist Mihaly Csikszentmihalyi:* Csikszentmihalyi's works are most easily accessible in his book *Flow* (1990), and, unless otherwise noted, the quotes in this chapter have been taken from pages 162 and 164 of that book.

33 *Brenda Laurel's analysis of "first-person" experience and Susanne Bødker's "human activity approach":* I discovered Brenda Laurel's work many years ago while leading a research project on the design of computer systems. We invited her to join us and to provide a chapter for our book (Laurel, 1986). This became the first publication of her work on first-person and third-person experiences. Her most recent book (Laurel, 1991: *Computer as Theatre*) shows how concepts from drama can make valuable contributions toward the construction of computer (and other) systems that can lead to an engaging experience. Laurel doesn't discuss Csikszentmihalyi's work, but I believe the two are highly related.

Bødker belongs to the "Scandinavian school of design," one that emphasizes the continual interaction of workers and designers. Bødker has investigated the nature of the work interaction and the "activity flow" during a work session, again a topic I think related to Csikszentmihalyi's. See her 1989 paper and her more comprehensive 1990 book (Bødker, 1989, 1990). In addition, her more recent work with Liam Bannon is highly relevant: Bannon and Bødker (1991).

34 *Who knows better how to sustain interaction and interest than those who create theatrical experiences?:* See Laurel's *Computer as Theatre* (1991). Some of this paragraph, I must admit, is taken from my foreword to her book.

34 *What is needed is "direct engagement," the feeling of directly working on the task:* See Laurel (1986) and Hutchins, Hollan, and Norman (1986).

35 *"Using a pager on the teacher's desk":* Csikszentmihalyi, in Greenberger (1992, p. 32).

38 *"Recent developments in computerized interactive multimedia can take us considerably further":* Leonard (1992, p. 28).

THREE

THE POWER OF REPRESENTATION

Page **Notes**

45 *SOCRATES: Then anyone who leaves behind him a written manual:* Plato (1961), *Plato: Collected Dialogues* (from the chapter "Phaedrus," a dialogue between Socrates and Phaedrus, p. 276). Phaedrus seems to be a wimp as a debating partner. It isn't clear why having a debating partner who responds "Very true" or "Once again you are perfectly right" after each interchange adds anything to the words that have just been spoken—they might just as well have been read (as indeed, you have just done). If I had graduate students who behaved that way, I would kick them out of the program.

46 *Reading was generally done aloud:* Noakes (1988).

46 *"Today, many readers take as the hallmark of the good novel":* Noakes.

46 *Readers were taught the rules of rhetoric:* Paraphrased from Noakes.

49 *A representational system has two essential ingredients:* Actually, there is a third component: an interpreter that keeps track of, interprets, and operates upon the relationships between the symbols in the representing world and the concepts and objects in the represented world. For the purposes of this book, we do not need to discuss the interpreter, but something has to maintain the relationship between the symbols in

the representation and the objects and concepts being referred to. In some kinds of perceptual representations, the interpretive property is embedded in the form of the representation and the machinery of the mind.

53 *"Solving a problem simply means representing it so as to make the solution transparent":* The quotation comes from Herbert Simon's classic book *The Sciences of the Artificial* (1981, p. 153). Also see Kaplan and Simon, "In Search of Insight" (1990).

55 *The two games are what we call "problem isomorphs":* Herbert Simon introduced me to the entire domain of problem isomorphs and to the comparison between ticktacktoe and the game of 15. Simon has long emphasized the role of representation in problem solving. As the quotation opening this section reveals, he has stated that the key to problem solving is to find the representation that makes the answer transparent. His book *The Sciences of the Artificial* is the classic primer and advanced text (yes, both) in this field (Simon, 1981).

57 *There is a seven-hour time difference:* Normally, there is an eight-hour time difference, but when I first developed this example, the conflicting periods during which daylight savings time in California and summertime in England were in effect made it a seven-hour difference.

58 *Stephen Casner has shown:* My discussion of the various formats for presenting airline scheduling information was inspired by the work of Casner (1990), who used this (and other) examples to illustrate design principles. Casner adds even more information to the graphic depiction of airline schedules by letting the thickness of the line represent cost, and so on.

60 *And that let us move among all the formats:* After I had written this, I discovered a company marketing electronic access to the OAG, displaying the results in a variety of formats, including one graphic rendition not unlike the one I show, with the duration of each flight indicated by the length of an airplane fuselage (displaying an unnecessary bit of what Tufte [1983; also see Chapter 4] would call "chartjunk").

62 *How well do people cope with their prescriptions?:* The studies of "pill organizers" and the statistics on patient error in following prescriptions come from Park, Morrell, Frieske, Blackburn, and Birchmore (1991).

63 *From the work of psychologist Ruth Day:* Day covers many other representational issues in her paper "Alternative Representations" (Day, 1988).

66 *In Roman numerals, . . . it doesn't even matter in what order you write the symbols:* In early Roman numerals, although order technically did not matter, it was so inconvenient to use mixed order that the symbols were always written in descending order of value, with large symbols on the left. After a while, shortcuts crept into the writing. Instead of writing *40* as "XXXX," it was easier to write it as "XL": 10 less than 50, signified by putting the *X* for "10" just to the left of the *L* for "50." With this subtractive notation, widely used today, order does matter.

68 *For small numerical differences the length of the representation does not provide any information about its value:* The length of the representation for Arabic numbers is proportional to the logarithm of the value, but it only changes each time the range of values increases by a factor of ten. Thus the representation for *23* is not longer than that for *13,* but the representations for *230* and *2,300* are.

72 *An important design principle—naturalness:* The "naturalness" principle is described in more technical language in Norman (1991). Naturalness is actually specified as the complexity of the description of the mapping between the representation and the thing being represented. This is a way of specifying what I called a "natural mapping" in *The Design of Everyday Things* (Norman, 1990).

FOUR

FITTING THE ARTIFACT TO THE PERSON

Page **Notes**

84 *One of my former graduate students Jiajie Zhang.* Zhang (1991, 1992).

92 *Psychologists have determined that perceived intensity of light and sound follows roughly a cube-root law:* I refer here primarily to the pioneering work of S. S. Stevens, who demonstrated that many additive domains follow a power law (now usually called "Steven's law"), in that psychological judgment (J), is given by physical intensity (I) raised to some power, (p). J

is proportional to I^p. In the case of loudness and brightness, the power is 0.3 (Stevens, 1957, 1975; Bolanowski, Stanley, and Gescheider, 1974; Moskowitz, Scharf, and Stevens, 1974).

93 *Graphic presentations were not used by American businesses until the late 1800s and early 1900s:* From Yates (1989, p. 85), who gives a major treatment of these issues throughout her book.

95 *It is estimated that Tokyo will be five times as large as Beijing:* From *The World Almanac and Book of Facts* (New York: World Almanac [Pharos Books], 1988, p. 739).

97 *There are numerous psychological principles that can be used to guide the construction of appropriate graphic relationships:* I highly recommend four major works on the display of information: Tufte's two books, *The Visual Display of Quantitative Information* (1983) and *Envisioning Information* (1990), both informative and beautiful to behold, books that follow their own principles; Cleveland's *The Elements of Graphing Data* (1985) is the third standard reference; Bertin's *Semiology of Graphics* (1983) is a bible of insight into the relationship between the psychological interpretation of information and the method of presentation (but it is almost incomprehensible—if you read this book, expect to work hard). I enjoyed Tufte's two books the most, but I learned more from Bertin.

My complaint about Bertin, Cleveland, and Tufte is that they all try to prescribe the "proper" graph. To me, there is no such thing divorced from the usage. The task dictates what kind of information should be displayed, how much, and in what form. It is not possible to speak of the method of presentation without also speaking of the use to which the information will be put.

"Chartjunk," is what Edward Tufte (1983) calls many examples of charts and graphs. "Too many data presentations, alas, seek to attract and divert attention by means of display apparatus and ornament. Chartjunk has come to corrupt all sorts of information exhibits and computer interfaces." Chartjunk isn't all bad, however. The use of symbols and pictures does help motivation. More important, it can provide mnemonic value, making it easier for the reader to remember what each line or symbol of the chart or graph stands for. Here the instincts of the graphics designer and the statistician differ.

97 *Appropriateness principle:* This principle was suggested by the work of Card, Mackinlay, and Robertson, "A Morphological Analysis of the Design Spaces of Input Devices" (1991). Mackinlay's Ph.D. thesis (1986) at Stanford first showed me how psychological principles could be applied to the understanding of graphs. (See Figure 4.3.)

106 *One message at a time:* True, one can get television sets that have a picture in the picture (PIP, it's called), allowing the viewer to watch two channels at the same time. I have seen as many as sixteen channels displayed at once on one screen. This doesn't change things, however. We can only concentrate on one of those channels at a time, and when the commercial comes, it is right there, where we are looking, where we can't escape it, at least not the first few seconds. One reason is that the commercial and the show being watched occupy the same spatial location.

FIVE

THE HUMAN MIND

Page **Notes**

117 *"The most complex structure in the known universe": Scientific American, 267,* no. 3 (September 1992), p. 4.

121 *In an important book, the psychologist Mervin Donald:* Donald's book is *Origins of the Modern Mind: Three Stages in the Evolution of Culture and Cognition* (1991).

123 *A considerable part of each day is spent establishing and maintaining social roles:* From Shirley Strum's studies of baboons in Kenya, reported in her chapter in Latour and Lemonnier (in press).

129 *"People like to tell stories," says Roger Schank:* Schank (1982, 1990) worried about the organizational structure of human memory and proposed a possible mechanism. His 1982 book focused on the problem of reminding, how it is that one experience reminds people of a previous one. The 1990 book, *Tell Me a Story,* expands the notion to a more organized form of information storage: stories. Schank suggests that stories are such a ubiquitous human experience because they are efficient memory-storage and organizational routines.

130 *Logical analysis only applies to information that can be measured:* I am aware that there are different kinds of logic, some of which try to overcome this limitation through the use of probabilities or other numerical scales that indicate degrees of uncertainty, "fuzzy" logic being the most popular attempt to get out of the binary, truth-falsity straitjacket. I commend the attempts, but the major critique is unchanged: The situation must still be abstracted to a bare resemblance of its actual self, and the logic can still only apply to things that are measurable, which is not the same as things that are important. I return to these issues in Chapter 9.

130 *There is no way to translate them into the language of logic, no way without badly distorting their content:* Modern decision making does try to put numerical values on these things, through rating scales, through translation into utility units or into some kind of equivalent monetary value. (For example, there is a large business in assessing the value of a human life in court cases.) I stand by my statement.

131 *"To err is human":* The best overall treatment of human error is James Reason's book *Human Error* (1990).

135 *Submitted the report to NASA's Aviation Safety Reporting System:* This incident was reported in NASA's monthly newsletter, *Callback*, for September 1991.

SIX

DISTRIBUTED COGNITION

Page **Notes**

142 *Without any need for talking:* Leon Segal's superb analysis of nonverbal cues in a helicopter cockpit first showed me the pertinence of this means of communication (Segal, 1990).

143 *Hutchins studied the navigation procedures:* Actually, Ed Hutchins has been my collaborator in a long-running series of studies, and many of my observations about the airline cockpit have come about through this collaboration. Although trained as a cognitive anthropologist, he is also now a licensed pilot who has taken the Boeing 747 flight courses and flown "jump seat" in the cockpit of many a commercial flight. His insights

into the nature of distributed cognition are of extreme importance to me and to the field: See Hutchins (in preparation).

145 *Automating factory control:* A number of studies concentrate on the social consequences of automation. Zuboff's book *In the Age of the Smart Machine: The Future of Work and Power* (1988) provides an important analysis of the impact of automation on plant performance. She shows how the same automation can have very different consequences in different plants, depending upon just how it was implemented and just what social structures were set up.

148 *We need to respond to the situation:* This is the crux of a new and controversial approach to the study of cognition called "situated action" (see Suchman, 1987, and Lave, 1988). I recommend the debate in the journal *Cognitive Science.* Started by the paper by Vera and Simon (1993: "Situated Action: A Symbolic Interpretation"), it led to spirited rebuttals by Greeno and Moore (1993), Agre (1993), Suchman (1993), and Clancey (1993). I was asked to be the "neutral referee" and to provide the introduction to the special issue of the journal in which the debate appeared.

148 *In the real world, it is not possible to do actions that are not possible:* This insight comes from Edwin Hutchins.

SEVEN

A PLACE FOR EVERYTHING, AND EVERYTHING IN ITS PLACE

Page **Notes**

159 *It is difficult today to realize that the common, everyday office filing cabinet represented a major revolution:* My knowledge of the history of the development of organizational filing systems comes from JoAnne Yates of MIT's Sloan School of Management, especially her book on the rise of organizational systems in business, which traces the history of information overload in the 1800s (Yates, 1989). During that period, advances in travel and communications gave rise to voluminous masses of materials, especially as business slowly discovered the importance of statistical record keeping. Her book traces the development of the letter press book, horizontal and then vertical filing cabinets, graphs, and then finally, the two

essential components of modern business: memos and the corporate newsletter.

168 *McGuckin's Hardware Store, in Boulder, Colorado:* The work of Brent Reeves and Gerhard Fischer of the Department of Computer Science at the University of Colorado introduced me to this fascinating and most useful store (Fischer and Reeves, 1992; Reeves, 1990).

170 *Why are dictionaries organized the way they are?:* McArthur (1986) provides an excellent history of reference books—encyclopedia, thesaurus, and dictionary—and the several attempts to overcome the alphabetical arrangement.

171 *Using an alphabet or a syllabary:* In an alphabet, each symbol (roughly) represents a consonant or vowel sound. To represent a syllable requires two or more alphabetic letters. Moreover, in many languages, the letters are pronounced differently depending upon the surrounding characters and even the meaning of the word being represented. Thus the alphabet is an imprecise representation of the spoken sound of a word. In a syllabary, each symbol represents a syllable. This makes for a much more precise translation between the printed word and its pronunciation. The same alphabet, however, can serve many languages, even those with different syllable structures.

173 *One of the powers of the computer should be that it doesn't have to keep things in order:* Actually, today's machines are serial in nature, doing operations only one step at a time. If they have a large amount of information to examine, then unless the information is properly organized, with good indexes, search can be slow. This is why the emphasis on sorting. What is needed is some sort of content-addressable storage medium. This can be approximated today by a technique known as "hash coding," but this only provides a partial solution. Neural networks, holographic memories, and parallel computers offer the hope of the more fundamental change that I am advocating. Things will change as these systems become more readily available. The goal I am seeking will be achieved, but not for a while.

184 *Paul Saffo, writing of what he called "our privacy jitters":* P. Saffo, "Future Tense: Personal Computers Will Make Solitude a Scarce Resource," *InfoWorld* 13, no. 51 (December 23, 1991), p. 37.

EIGHT

PREDICTING THE FUTURE

Page	**Notes**
185	*"Almost everything that has happened, and its opposite, has been prophesied":* The quotation from Herbert Simon was originally published in 1977, but I took it from the reprinting of his article in the collection of readings edited by Pylyshyn and Bannon (1989, p. 445).
185	*We do know that new technologies will bring both dividends and problems, especially human, social problems:* Studies of the history of technology have emphasized the technology and the social and political impacts but neglected the cognitive impact on the individual. This probably reflects the historian's trend toward the study of societies, not of individuals. The major exception is studies of the impact of writing and printing. The same neglect of cognition can be found in predictions regarding the future of technology. Here, once more, the emphasis is on the technology itself and perhaps its most immediate, first-order impact, with very little thought given to the manner by which it might change individual cognition. Science fiction authors stand out as the group who have thought most about these factors, except that they often focus upon the negative side of technology, how it can lead to a mass of disenfranchised, dissatisfied inhabitants.
186	*A humbling way to begin a look toward the future is to examine the past:* My information about past predictions comes from a wide variety of places. Most fascinating are two books by Corn: *Imagining Tomorrow: History, Technology, and the American Future* (Corn, 1986) and the museum catalog and collection of photographs and drawings *Yesterday's Tomorrows: Past Visions of the American Future* (Corn and Horrigan, 1984). Marvin's (1988) *When Old Technologies Were New* continues in a similar vein, discussing in detail the growth of electric communication in the late nineteenth century. My information about the early days of the computer comes from Ceruzzi's (1986) chapter in Corn (1986), from Augarten's (1984) review, and from my own experiences as a programmer of the Univac I computer. I also used *Panati's Browser's Book of Beginnings* (Panati, 1984) for some dates. I have not provided a detailed history of technology—nor

am I the person to do so. Some useful secondary sources are Hughes (1989) and Boorstin (1983). Latour (1986, 1987) provides a discussion of the impact of social forces upon science. The classic treatment of the mechanization of work is by Giedion (1948): *Mechanization Takes Command: A Contribution to Anonymous History.* The book is out of print and difficult to find, but it is well worth the effort.

188 *The plants are too complex to be run:* See Perrow's *Normal Accidents* (1984). In addition, I have worked in nuclear power safety, and some of these comments come from my own observations. I think many will agree that the failure of nuclear power is a result of the social side, not the technology. Except that proponents of nuclear power will say it damningly, stating that those nontechnological people ruined it all, still insisting that the plants are safe and that occasional accidents are a result of human failure, human error. In my essay "Coffee Cups in the Cockpit," I called this the "train and blame" philosophy (Norman, 1992). And opponents will say, "Yeah, see." I am trying to be somewhere in between (and thereby liked by neither side?).

190 *How society would modify the original notion:* Bijker, Hughes, and Pinch's (1987) *The Social Construction of Technological Systems* presents an analysis of the manner by which society influences the forms and directions of technology—just the opposite of the normal direction of analysis.

190 *The telephone:* My quotations about the early uses (and abuses) of the telephone come from Marvin (1988).

194 *For a technology to be available to the home or business in ten years, there must now be working prototypes:* I am indebted to Bill Curtis for this observation about technological development.

198 *John Seely Brown has pointed out that one of the things that binds together a culture is that everyone reads the same newspapers, sees the same shows:* In his address at the 1992 annual conference of the Association for Computing Machinery's special interest group on Computer-Human Interaction (called CHI-92).

204 *We already have at least one computer program that generates art:* Harold Cohen's program Aaron. (McCorduck, 1991). Note that museums insist that each exhibition include both

"Cohen" and "Aaron." And Cohen's initial desire that the paintings be signed only by the computer has been thwarted by the art world's insistence that he, Harold Cohen, sign each one (thereby maintaining both responsibility and credit for the work).

206 *Vernor Vinge wrote "True Names":* Vinge's story is available in his collection (Vinge, 1984). Virtual reality is now a popular research topic, and discussions can be found in much of the popular press. Some technical observations are presented in Laurel (1990; see the photographs in the center of the book and the article by Fisher and Ellis in that book). My longtime favorite for pure art and enjoyment is Krueger's *Artificial Reality*, a most enjoyable experience (1991).

210 *Science Fiction and . . . :* Taken from a talk I originally presented at the CHI-93 Conference on Computer-Human Interaction, Monterey, California, May 1992, for a symposium on science fiction and the design of computer systems. The symposium was organized by the graphics designer Aaron Marcus with science fiction writers Rudy Rucker, Bruce Sterling, and Vernor Vinge, all of whom I must thank for inspiring this section.

213 *A pianist playing a Chopin nocturne:* The nocturne I had in mind is F. Chopin's Opus 5, No. 1, Nocturne in F. This and the next sentence are taken from my book *Memory and Attention* (Norman, 1976, pp. 200–201).

215 *CSCW: computer-supported cooperative work:* Artifacts that aid group work are described in a number of places. I find Lakin's (1985) description of a "performance aid" for group work particularly insightful: a low-technology, but powerful aid. His paper can be found in the collection of readings on early work to support group efforts (Greif, 1988); this collection is an excellent place to begin. Also see Schrage (1990) for a very readable review of ongoing research in support of joint work. The journal *Communications of the ACM* devoted a special issue to cooperative work (in January 1991). The discussion of the overlapping video workplace where tutor and student could work together is reported there (Ishii and Miyake, 1991). Some of the work I discuss in this section comes from my knowledge of the work done at Bell Communication Research Corporation (Bellcore) and at Xerox Corporation's Palo Alto Research Center (PARC) and the branch of

PARC in Cambridge, England (EuroPARC). "Cruiser" (Root, 1988) was the video system I had in mind that allowed people to electronically "cruise the halls," visiting colleagues in their offices. Electronic conference rooms are by now commonplace, existing in research laboratories throughout industry and academia; in some industries, they are in daily use.

The yearly proceedings of the Human Factors in Computing Systems Conferences *(CHI-90, CHI-91, CHI-92, etc.)* and the proceedings of the CSCW conferences are filled with research reports on all these topics. These are the places to look for the latest research. (The CHI proceedings are published by the Association for Computing Machinery [ACM] and are available from Addison-Wesley Publishing Company.)

216 *Here is a social tradeoff:* In Chapter 4, I call this "Grudin's law," after Jonathan Grudin, who has done considerable research on the social implications of new technologies and who first formulated this tradeoff. His paper "Social Evaluation of the User Interface: Who Does the Work and Who Gets the Benefit" (Grudin, 1987) was especially influential. You should also see his 1989 paper on "groupware" (another name for "computer-supported cooperative work").

NINE

SOFT AND HARD TECHNOLOGY

Page **Notes**

223 *In 75 percent of commercial aviation accidents, the blame is put on the pilots:* From the United States National Transportation Safety Board's *Annual Review of Aircraft Accident Data: U.S. Air Carrier Operations Calendar Year 1987* (NTSB, 1990).

225 *Automation works best when conditions are normal:* Wiener (1988), Wiener and Curry (1980).

235 *People could use nicknames or pseudonyms:* Calling number identification can be used to block crank, obscene, and other unwanted phone calls, if only because these callers know that you know their number and therefore can give it to the police or other authorities. My system of person identification lacks this feature. Moreover, the obscene phone caller will have no scruples about lying during the identification phase. The tech-

nology of number identification provides a remedy: One could add a "call-blocking" service that prevented calls from specific calling numbers from being received, but without revealing the numbers (this service is already offered in some locations). This way, an obscene caller could get through the first time, but not thereafter. A similar service would allow the recipient to tag a telephone call so that the telephone company or other permissible agency could determine the number of the calling party and track down the caller. This service also already exists in some locations. Both these services seem to provide appropriate protection against unwanted and harassing calls, without violating privacy.

237 *The computer newsletters to which I subscribe: RISKS* is an electronic newsletter dedicated to the study of computer (and other) risks in society. It is run by Peter Neumann and sponsored by the Association for Computing Machinery, the professional society for American computer scientists, and is available on the internet "netnews" as comp.risks. Summaries are published each month in the journal *Communications of the ACM*.

240 *Rabbit:* The "Rabbit" interface for information retrieval is described in Tou (1982); Williams (1984); and Williams, Tou, Fikes, Henderson, and Malone (1982).

TEN

TECHNOLOGY IS NOT NEUTRAL

Page **Notes**

243 *"The medium is the message":* McLuhan (1964).

244 *Remember what Socrates, the great Greek philosopher, said:* See page 45 (Chapter 3).

245 *As Jerry Mander put it:* Mander (1991, pp. 76 and 81).

247 *A compositional medium:* Newell (1991).

247 *The Shakespeare Project:* Developed by Larry Friedlander in the English department at Stanford University. Actually, there are two different programs developed jointly by Friedlander and the Faculty Author Development Program at Stanford: "The Shakespeare Project" and "The Theater Game." The Shakespeare Project comes with computer program and laser-

disc video recordings of different acting groups performing *Hamlet.* Viewers analyze the scenes by writing their interpretations of the inner thoughts of the characters, by identifying the "beats" of the enacted scenes, and doing careful, critical analysis of the varied interpretations shown on the videodisc. They can then compare their analyses with those of the instructors.

In The Theater Game, students interpret Shakespeare with the help of the computer. One of the program's developers, Charles Kerns, described it to me this way: "This was done on a 2½–dimension animation system that let students block a scene, design the stage and props, select costumes, and then move the characters to their appointed places as the scene progressed. One could have Hamlet face in different directions, turn his head, sit, kneel, stand, and importantly, fall dead on the ground. The combination of the interpretative and generative aspects of the program seemed one of its strengths. Widespread close study of visual texts seems like one of the interesting fallouts of this technology" (Kerns, electronic mail, 1991).

249 *Am I not succumbing to the "everything is OK" philosophy?:* Mander (1991, pp. 2–3).

250 *"Guilt," say Kubey and Csikszentmihalyi:* Kubey and Csikszentmihalyi (1990).

252 *In his book* Amusing Ourselves to Death: Postman (1985). The quotation comes from page 5 of the Voyager electronic book edition.

REFERENCES

Agre, P. E. (1993). The symbolic world view. Reply to Vera and Simon. *Cognitive Science*. In press.

Augarten, S. (1984). *Bit by bit: An illustrated history of computers*. London: Unwin Paperbacks.

Bannon, L. J. and Bødker, S. (1991). Beyond the interface: Encountering artifacts in use. In J. M. Carroll (Ed.), *Designing interaction: Psychology at the human-computer interface* (pp. 227–253). New York: Cambridge University Press.

Bertin, J. (1983). *Semiology of graphics* (William J. Berg, Trans.). Madison, WI: University of Wisconsin Press.

Bijker, W. E., Hughes, T. P., and Pinch, T. J. (Eds.). (1987). *The social construction of technological systems*. Cambridge, MA: MIT Press.

Bødker, S. (1989). A human activity approach to user interfaces. *Human-Computer Interaction, 4*, 171–195.

Bødker, S. (1990). *Through the interface: A human activity approach to user interface design*. Hillsdale, NJ: Erlbaum.

Bolanowski, J., Stanley, J. and Gescheider, G. A. (Eds.). (1974). *Ratio scaling of psychological magnitude: In honor of the memory of S. S. Stevens*. Hillsdale, NJ: Erlbaum.

Boorstin, D. J. (1983). *The discoverers*. New York: Random House. (Paperback edition published by Vintage Books, 1983.)

Bruner, J. S. (1986). *Actual minds, possible worlds*. Cambridge, MA: Harvard University Press.

Card, S. K., Mackinlay, J. D., and Robertson, G. G. (1991). A morphological analysis of the design spaces of input devices. *ACM Transactions on Information Systems, 9*, 99–122.

Casner, S. M. (1990). *Task-analytic design of graphic presentations.* Unpublished Ph.D. dissertation, Intelligent Systems Program, University of Pittsburgh.

Ceruzzi, P. (1986). An unforeseen revolution: Computers and expectations, 1935–1985. In J. J. Corn (Ed.), *Imagining tomorrow: History, technology, and the American future* (pp. 188–201). Cambridge, MA: MIT Press.

Cheney, D. L., and Seyfarth, R. M. (1990). *How monkeys see the world: Inside the mind of another species.* Chicago: University of Chicago Press.

Clancey, W. J. (1993). Situated action: A neuropsychological interpretation. *Cognitive Science.* In press.

Cleveland, W. S. (1985). *The elements of graphing data.* Monterey, CA: Wadsworth.

Corn, J. J. (Ed.). (1986). *Imagining tomorrow: History, technology, and the American future.* Cambridge, MA: MIT Press.

Corn, J. J., and Horrigan, B. H. (1984). *Yesterday's tomorrows: Past visions of the American future.* (Smithsonian Institution Travelling Exhibition Service, Washington ed.) New York: Summit Books.

Csikszentmihalyi, M. (1990). *Flow: The psychology of optimal experience.* New York: Harper & Row.

Day, R. S. (1988). Alternative representations. In G. Bower (Ed.), *The psychology of learning and motivation* (pp. 261–305). Orlando, FL: Academic Press.

Donald, M. (1991). *Origins of the modern mind: Three stages in the evolution of culture and cognition.* Cambridge, MA: Harvard University Press.

Fischer, G., and Reeves, N. B. (1992). Beyond intelligent interfaces: exploring, analyzing and creating success models of cooperative problem solving. *Applied Intelligence, special issue intelligent interfaces,* 1, 311–332.

Giedion, S. (1948). *Mechanization takes command: A contribution to anonymous history.* Oxford, UK: Oxford University Press. Republished in 1969 by Norton (New York).

Greenberger, M. (Ed.). (1992). *Technologies for the 21st century.* Volume 3: *Multimedia in review.* Santa Monica, CA: Voyager.

Greeno, J., and Moore, J. L. (1993). Situativity and symbols: Response to Vera and Simon. *Cognitive Science.* In press.

Gregory, R. L. (1989, March 25). Can hands-on exploration turn minds on to science? Unpublished report.

Greif, I. (1988). *Computer-supported cooperative work: A book of readings*. San Mateo, CA: Morgan Kaufmann.

Grudin, J. (1987). Social evaluation of the user interface: Who does the work and who gets the benefit. *Proceedings of INTERACT 87: IFIF Conference on Human-Computer Interaction*. (Stuttgart, Germany.)

Grudin, J. (1989). Why groupware fails: Problems in design and evaluation. *Office Technology and People, 4*, 245–264.

Hughes, T. P. (1989). *American genesis: A century of invention and technological enthusiasm, 1870—1970*. New York: Viking Penguin.

Hutchins, E. (In preparation). *Distributed cognition*.

Hutchins, E.; Hollan, J.; and Norman, D. A. (1986). Direct manipulation interfaces. In D. A. Norman and S. Draper (Eds.), *User centered system design: New perspectives in human-computer interaction*. Hillsdale, NJ: Erlbaum.

Huxley, A. (1932). *Brave new world*. London: Chatto & Windus.

Huxley, A. (1953). *Brave new world revisited*. New York: Harper & Row.

Ishii, H., and Miyake, N. (1991). TeamWorkStation: Towards an open shared workspace. *Communications of the ACM, 34*, 37–50.

Jolly, A. (1991). Thinking like a vervet. (Review of Cheney and Seyfarth [1990], *How monkeys see the world: Inside the mind of another species.) Science, 251*, 574–557.

Kaplan, C. A., and Simon, H. A. (1990). In search of insight. *Cognitive Psychology, 22*, 374–419.

Krueger, M. W. (1991). *Artificial reality* (2nd ed.). Reading, MA: Addison-Wesley.

Kubey, R., and Csikszentmihalyi, M. (1990). *Television and the quality of life: How viewing shapes everyday experience*. Hillsdale, NJ: Erlbaum.

Lakin, F. (1990). Visual language for cooperation: A performing medium approach for systems for cooperative work. In J. Galegher, R. E. Kraut, and C. Egido (Eds.), *Intellectual teamwork: Social and technical bases of collaborative work* (pp. 453–488). Hillsdale, NJ: Erlbaum.

Latour, B., and Lemonnier, P. (Eds.). (1993). *L'intelligence social des techniques*. Paris: La Decouverte. (In press).

Latour, B. (1986). Visualization and cognition: Thinking with eyes and hands. *Knowledge and Society: Studies in the Sociology of Culture Past and Present, 6*, 1–40.

Latour, B. (1987). *Science in action*. Cambridge, MA: Harvard University Press.

Laurel, B. (1991). *Computer as theatre: A dramatic theory of interactive experience*. Reading, MA: Addison-Wesley.

Laurel, B. (Ed.). (1990). *The art of human-computer interface design*. Reading, MA: Addison-Wesley.

Laurel, B. K. (1986). Interface as mimesis. In D. A. Norman and S. W. Draper (Eds.), *User centered system design* (pp. 67–85). Hillsdale, NJ: Erlbaum.

Lave, J. (1988). *Cognition in practice*. Cambridge, UK: Cambridge University Press.

Leonard, G. (1992, May). The end of school. *The Atlantic*, 269 (5), 24–32.

McArthur, T. (1986). *Worlds of reference*. Cambridge, UK: Cambridge University Press.

Mackinlay, J. D. (1986). Automating the design of graphical presentations of relational information. *ACM Transactions on Graphics*, 5(2), 110–141.

McCorduck, P. (1991). *Aaron's code: Meta-art, artificial intelligence, and the work of Harold Cohen*. New York: Freeman.

McLuhan, M. (1964). *Understanding media*. New York: McGraw-Hill.

Mander, J. (1991). *In the absence of the sacred: The failure of technology and the survival of the Indian nations*. San Francisco: Sierra Club Books.

Marvin, C. (1988). *When old technologies were new: Thinking about electric communication in the late nineteenth century*. New York: Oxford University Press.

Moskowitz, H. R.; Scharf, B.; and Stevens., J. C. (Eds.). (1974). *Sensation and measurement: Papers in honor of S. S. Stevens*. Dordrecht, Netherlands: Reidel.

Mumford, L. (1934). *Technics and civilization*. New York: Harcourt Brace & World.

Mumford, L. (1938). *The culture of cities*. New York: Harcourt Brace & World.

nology of number identification provides a remedy: One could add a "call-blocking" service that prevented calls from specific calling numbers from being received, but without revealing the numbers (this service is already offered in some locations). This way, an obscene caller could get through the first time, but not thereafter. A similar service would allow the recipient to tag a telephone call so that the telephone company or other permissible agency could determine the number of the calling party and track down the caller. This service also already exists in some locations. Both these services seem to provide appropriate protection against unwanted and harassing calls, without violating privacy.

237 *The computer newsletters to which I subscribe: RISKS* is an electronic newsletter dedicated to the study of computer (and other) risks in society. It is run by Peter Neumann and sponsored by the Association for Computing Machinery, the professional society for American computer scientists, and is available on the internet "netnews" as comp.risks. Summaries are published each month in the journal *Communications of the ACM*.

240 *Rabbit:* The "Rabbit" interface for information retrieval is described in Tou (1982); Williams (1984); and Williams, Tou, Fikes, Henderson, and Malone (1982).

TEN

TECHNOLOGY IS NOT NEUTRAL

Page **Notes**

243 *"The medium is the message":* McLuhan (1964).

244 *Remember what Socrates, the great Greek philosopher, said:* See page 45 (Chapter 3).

245 *As Jerry Mander put it:* Mander (1991, pp. 76 and 81).

247 *A compositional medium:* Newell (1991).

247 *The Shakespeare Project:* Developed by Larry Friedlander in the English department at Stanford University. Actually, there are two different programs developed jointly by Friedlander and the Faculty Author Development Program at Stanford: "The Shakespeare Project" and "The Theater Game." The Shakespeare Project comes with computer program and laser-

disc video recordings of different acting groups performing *Hamlet.* Viewers analyze the scenes by writing their interpretations of the inner thoughts of the characters, by identifying the "beats" of the enacted scenes, and doing careful, critical analysis of the varied interpretations shown on the videodisc. They can then compare their analyses with those of the instructors.

In The Theater Game, students interpret Shakespeare with the help of the computer. One of the program's developers, Charles Kerns, described it to me this way: "This was done on a 2½–dimension animation system that let students block a scene, design the stage and props, select costumes, and then move the characters to their appointed places as the scene progressed. One could have Hamlet face in different directions, turn his head, sit, kneel, stand, and importantly, fall dead on the ground. The combination of the interpretative and generative aspects of the program seemed one of its strengths. Widespread close study of visual texts seems like one of the interesting fallouts of this technology" (Kerns, electronic mail, 1991).

249 *Am I not succumbing to the "everything is OK" philosophy?:* Mander (1991, pp. 2–3).

250 *"Guilt," say Kubey and Csikszentmihalyi:* Kubey and Csikszentmihalyi (1990).

252 *In his book* Amusing Ourselves to Death: Postman (1985). The quotation comes from page 5 of the Voyager electronic book edition.

Mumford, L. (1967). *The myth of the machine: Technics and human development.* New York: Harcourt Brace Jovanovich.

Newell, A. (1991). *Unified theories of cognition.* Cambridge, MA: Harvard University Press. (The 1987 William James Lectures at Harvard University.)

Noakes, S. (1988). *Timely reading: Between exegesis and interpretation.* Ithaca: Cornell University Press. (See Chapter 1: A sketch of a fragment from a story of reading, 14–37.)

Norman, D. A. (1976). *Memory and attention* (2nd ed.). New York: Wiley.

Norman, D. A. (1982). *Learning and memory.* New York: Freeman.

Norman, D. A. (1990). *The design of everyday things.* New York: Doubleday. (Originally published as *The psychology of everyday things.* New York: Basic Books, 1988.)

Norman, D. A. (1991). Cognitive artifacts. In J. M. Carroll (Ed.), *Designing interaction: Psychology at the human-computer interface* (pp. 17–38). New York: Cambridge University Press.

Norman, D. A. (1992). *Turn signals are the facial expressions of automobiles.* Reading, MA: Addison-Wesley.

Norman, D. A. (1993). Cognition in the head and in the world: Introduction to a debate on situated action. *Cognitive Science.* In press.

Norman, D. A., and Draper, S. W. (Eds.). (1986). *User centered system design.* Hillsdale, NJ: Erlbaum.

NTSB. (1990, November 29). *Annual review of aircraft accident data: U.S. air carrier operations calendar year 1987* (Report No. NTSB/ARC-90/01, Govt. Accession No. PB 91/119693). Washington, DC: National Transportation Safety Board.

Orwell, G. (1950). *1984.* New York: New American Library.

Panati, C. (1984). *Panati's browser's book of beginnings.* Boston: Houghton Mifflin.

Park, D. C.; Morrell, R. W.; Frieske, D.; Blackburn, A. B.; and Birchmore, D. (1991). Cognitive factors and the use of over-the-counter medication organizers by arthritis patients. *Human Factors, 33,* 57–67.

Perrow, C. (1984). *Normal accidents.* New York: Basic Books.

Plato. (1961). *Plato: Collected dialogues.* Princeton, NJ: Princeton University Press.

Postman, N. (1985). *Amusing ourselves to death: Public discourse in the age of show business.* New York: Viking Penguin.

Postman, N. (1992). *Technopoly.* New York: Knopf.

Pursell, C. W. (1979). Government and technology in the great depression. *Technology and Culture, 20*(1). (Also see Pursell [1990].)

Pursell, C. W. (Ed.). (1990). *Technology in America: A history of individuals and ideas.* Cambridge, MA: MIT Press.

Pylyshyn, Z. W., and Bannon, L. J. (Eds.). (1989). *Perspectives on the computer revolution.* Norwood, NJ: Ablex.

Quin, M. (1991, October 26). All grown up and ready to play: Melanie Quin celebrates the role of interactive science centres. *New Scientist,* 60–61.

Rabinbach, A. (1990). *The human motor: Energy, fatigue, and the origins of modernity.* New York: Basic Books.

Rasmussen, J. (1983). Skills, rules, and knowledge: Signals, signs, and symbols and other distinctions in human performance models. *IEEE Transactions on Systems, Man, and Cybernetics, 13,* 257–266.

Reason, J. (1990). *Human error.* Cambridge, UK: Cambridge University Press.

Reeves, B. (1990). *Finding and choosing the right object in a large hardware store—An empirical study of cooperative problem solving among humans.* (Technical report.) Department of Computer Science, University of Colorado, Boulder.

Root, R. W. (1988). Design of a multi-media vehicle for social browsing. In *Proceedings of ACM CSCW'88 Conference on Computer-Supported Cooperative Work* (pp. 25–38).

Rumelhart, D. E., and Norman, D. A. (1978). Accretion, tuning and restructuring: Three modes of learning. In J. W. Cotton and R. Klatzky (Eds.), *Semantic factors in cognition.* Hillsdale, NJ: Erlbaum.

Rumelhart, D. E., and Norman, D. A. (1981). Analogical processes in learning. In J. R. Anderson (Ed.), *Cognitive skills and their acquisition.* Hillsdale, NJ: Erlbaum.

Saffo, P. (1991). Future tense: Personal computers will make solitude a scarce resource. *InfoWorld,* 13, no. 51 (December 23).

Schank, R. C. (1982). *Dynamic memory.* New York: Cambridge University Press.

Schank, R. C. (1990). *Tell me a story: A new look at real and artificial memory.* New York: Scribner.

Schrage, M. (1990). *Shared minds: The new technologies of collaboration.* New York: Random House.

Segal, L. D. (1990). Effects of aircraft cockpit design on crew communication. In E. J. Lovesey (Ed.), *Contemporary ergonomics 1990.* UK: Taylor & Francis.

Shastri, L., and Ajjanagadde, V. (1993). A connectionist model of reflexive processing. *Behavioral and Brain Sciences.* In press.

Showalter, J. C., and Driesbach, J. (Eds.). (1983). *Wooton patent desks: A place for everything and everything in its place.* Indianapolis: Indiana State Museum; also Oakland, CA: The Oakland Museum. (Publication to accompany a museum exhibition organized by the Indiana State Museum and The Oakland Museum.)

Simon, H. A. (1981). *The sciences of the artificial* (2nd ed.). Cambridge, MA: MIT Press.

Stevens, S. S. (1957). On the psychophysical law. *Psychological Review, 64,* 153–181.

Stevens, S. S. (1975). *Psychophysics: Introduction to its perceptual, neural, and social prospects.* New York: Wiley. (Reprinted in 1986. New Brunswick, NJ: Transaction Books. The 1986 edition, edited by Geraldine Stevens, has a new introduction by Lawrence E. Marks.)

Suchman, L. (1987). *Plans and situated actions: The problem of human-machine communication.* New York: Cambridge University Press.

Suchman, L. (1993). Situativity and symbols. Response to Vera and Simon. *Cognitive Science.* In press.

Tou, F. N. (1982). *Rabbit: A novel approach to information retrieval.* Unpublished master's thesis, Massachusetts Institute of Technology, Cambridge.

Tufte, E. R. (1983). *The visual display of quantitative information.* Cheshire, CT: Graphics Press.

Tufte, E. R. (1990). *Envisioning information.* Cheshire, CT: Graphics Press.

Vera, A., and Simon, H. (1993). Situated action: A symbolic interpretation. *Cognitive Science.* In press.

Vinge, V. (1984). *True names.* New York: Bluejay Books (distributed by St. Martin's Press).

Wiener, E. L. (1988). Cockpit automation. In E. L. Wiener and D. C. Nagel (Eds.), *Human factors in aviation.* Orlando, FL: Academic Press.

Wiener, E. L., and Curry, R. E. (1980). Flight-deck automation: Promises and problems. *Ergonomics, 23,* 995–1011. (Also reprinted in R. Hurst and L. R. Hurst [Eds.]. *Pilot error: The human factor.* New York: Jason Aronson, 1982.)

Williams, M. (1984). What makes Rabbit run? *International Journal of Man-Machine Studies, 21,* 333–352.

Williams, M. D., Tou, F. N., Fikes, R. E., Henderson, D. A. and Malone, T. (1982). Rabbit: Cognitive science in interface design. *Proceedings of the Cognitive Science Society.* (Ann Arbor, MI) Hillsdale, NJ: Erlbaum.

Wymer, P. (1991, October 5). Never mind the science, feel the experience: Paul Wymer feels the time has come to question the purpose of interactive science displays. *New Scientist.*

Yates, J. (1989). *Control through communication: The rise of system in American management.* Baltimore: Johns Hopkins University Press.

Zhang, J. (1991). The interaction of internal and external representations in a problem solving task. *Proceedings of the Thirteenth Annual Conference of the Cognitive Science Society* (pp. 954–958; Chicago). Hillsdale, NJ: Erlbaum.

Zhang, J. (1992). *Distributed representation: The interaction between internal and external information.* Unpublished Ph.D. dissertation, Department of Cognitive Science, University of California, San Diego.

Zindell, D. (1989). *Neverness.* New York: Bantam.

Zuboff, S. (1988). *In the age of the smart machine: The future of work and power.* New York: Basic Books.

INDEX